The Future of Grain

Western grain was one of the keystones of Canada's National Policy, and the "Wheat Boom" early in this century propelled the economy into an unprecedented boom.

Today Western wheat occupies a much smaller share of the economy. Yet Canada's share of world exports of grains and oilseeds has picked up after a decline during the 1970s: Canada accounts for about 20 per cent of global wheat exports, making a healthy contribution to the country's balance of trade. The Canadian Wheat Board's goal is for exports to rise from the 1982-83 record level of 28 million tonnes to 36 million tonnes by 1990. How realistic is this objective? How can Canada reap the maximum benefit from expansion in wheat and other crops such as barley and rapeseed?

These are the questions that Terry and Michele Veeman examine in this study. For each crop in the grains and oilseeds sector, they provide a wealth of information on production, the transportation and handling system, exports, international competition, and the functions of government agencies.

The authors caution against over-optimism about the recent improvement in the fortunes of wheat. The all-important export market — some three-quarters of Canada's wheat goes abroad — is inherently unpredictable. The recent rise in exports was largely caused by external factors: drought in Australia, reduced U.S. exports to Russia and China, and relief from transportation bottlenecks in Canada due to the decline in rail traffic during the recession. However, the Veemans conclude that if global demand for wheat continues to rise, Canada can meet the Canadian Wheat Board goal — provided it upgrades the Prairie grain handling and rail system (as planned in the legislation that involves the end of the "Crow" rail rate) and invests in research and development.

The Veemans examine the prospects such an increase holds for the largely static Canadian industries connected to grain: manufacturers of agricultural machinery, pesticides and fertilizers; and processing industries such as feed and flour milling, rapeseed crushing and barley malting. And they suggest improvements to every aspect of government involvement — from measures to conserve Prairie soil to macroeconomic policy — that affects this important sector.

Michele and Terry Veeman are with the departments of rural economy and economics at the University of Alberta

The Canadian Institute for Economic Policy has been established to engage in public discussion of fiscal, industrial and other related public policies designed to strengthen Canada in a rapidly changing international environment.

The Institute fulfills this mandate by sponsoring and undertaking studies pertaining to the economy of Canada and disseminating such studies. Its intention is to contribute in an innovative way to the development of public policy in Canada.

Canadian Institute for Economic Policy
Suite 409, 350 Sparks St., Ottawa K1R 7S8

The Future of Grain

Canada's Prospects for Grains, Oilseeds and Related Industries

Terry Veeman and Michele Veeman

Canadian Institute for Economic Policy

The opinions expressed in this study are those of the authors alone and are not intended to represent those of any organization with which they may be associated.

ISBN 0-88862-622-3 paper
ISBN 0-88862-623-1 cloth

6 5 4 3 2 1 84 85 86 87 88 89

Canadian Cataloguing in Publication Data
Veeman, Terry.
The future of grain

1. Grain trade - Canada, Western. I. Veeman, Michele, 1941- II. Canadian Institute for Economic Policy. III. Title.

HD9044.C3V43 1984 338.1'731'09712 C84-098211-9

Additional copies of this book
may be purchased from:
James Lorimer & Company, Publishers
Egerton Ryerson Memorial Building
35 Britain Street
Toronto, Ontario M5A 1R7
Printed and bound in Canada

Contents

List of Tables and Figures

Foreword

The grain economy of Western Canada has been a mainstay of the Canadian economy and is expected to contribute substantially to our international balance of payments in future. However, there are substantial differences of opinion as to the extent of the future contribution of the grain sector to the economy.

Because of this uncertainty, the Institute commissioned a study to review the future of the grain economy within the context of a rapidly changing international economy environment and growing physical limitations on the Canadian resource base (eg., good land limits soil deterioration, increased salinity). The authors, Terry and Michele Veeman, have undertaken a very comprehensive review of this sector and have concluded that Canadian grain exports are expected to evidence moderate growth in the next ten years. They also review the implications of this growth scenario, which should provide a valuable input into the public discussion.

Like all our studies, the views expressed here are those of the authors and do not necessarily reflect those of the Institute.

Roger Voyer
Executive Director
Canadian Institute for Economic Policy

Acknowledgements

The original impetus to study the grains sector in Canada came from Dr. Roger Voyer, executive director of the Canadian Institute for Economic Policy, and Dr. Bruce Wilkinson, our colleague at the University of Alberta who is a director of the institute. Much of the recent debate on the grains sector in Canada has focused on grain transportation and revision of the historic Crowsnest Pass rates for grain. Our focus is somewhat different. Our main tasks were to assess the strength of possible expansion of Canadian grain production and exports, examine the potential impact this might have on sectors related to grain and the Canadian economy, and suggest policy improvements that might assist in the fuller realization of the growth opportunities associated with grains expansion.

We gratefully acknowledge the assistance of our colleagues at the University of Alberta who have cheerfully discussed grain matters over coffee for many months. We thank those who reviewed our manuscript for their helpful comments. We are especially indebted to Ellen Moreau who provided dedicated research assistance, particularly in compiling useful background material on the rapeseed and input supply sectors. Ellen also greatly helped with the compilation of tables and the generation of figures and with editing. We wish to thank Michael Austin for his research assistance in the early stages of the project. Barbara Johnson ably assisted with the documentation of references. The production of the manuscript was in the capable hands of Judy Warren, Clare Shier and Jim Copeland. We especially wish to thank Judy Warren for her many hours of textform entry and editing.

This study is dedicated to our children, Michael and Kathryn, and to the memory of Ted Veeman, a prairie grain farmer who in his lifetime saw both despair and promise in the grains sector.

The Grains Challenge 1

Grain, particularly wheat, has been a very important resource staple in the economic development of the prairie region and the Canadian economy. Over time, however, the relative role of the agricultural sector in the Canadian economy has considerably declined. Today, in the 1980s, there are prospects of moderate expansion in Canadian grain production and exports. To what degree can such expansion of this historically important resource staple provide growth opportunities in the prairie region and in the overall Canadian economy in the 1980s and 1990s?

After two decades of rapid expansion, world trade in grains and oilseeds declined in the early 1980s (Figure 1-1). During the 1960s and 1970s, Canadian exports of grains and oilseeds increased at a slower pace than either world trade in grains or American exports, leading to concern in Canada about our declining market share. However, in the recessionary times of the early 1980s, Canada has, paradoxically, achieved record level grain exports and increased its share of world wheat exports back to earlier levels of over 20 percent. Anticipating continued expansion in world trade in grains, the Canadian Wheat Board in 1980 set out grain export targets of 30 million tonnes (metric tons) by 1985 and 36 million tonnes by 1990. A major symposium on prairie grain production in the fall of 1980 concluded that the production levels (assumed at that time to be 54 million tonnes) needed to generate such export levels were technically possible. More recently, the Canada Grains Council, in a detailed study of production possibilities and system flows, has slightly revised these production and export goals and suggested that 50 million tonnes of western grain and oilseed production in 1990 are compatible with Canadian export levels of 34 million tonnes.

At first glance, given the short-run difficulties of the world economy and the relative stagnation of world grain trade in 1982 and 1983, it

FIGURE 1-1
EXPORTS OF MAJOR GRAINS AND OILSEEDS: WORLD, UNITED STATES AND CANADA, 1922 TO 1983

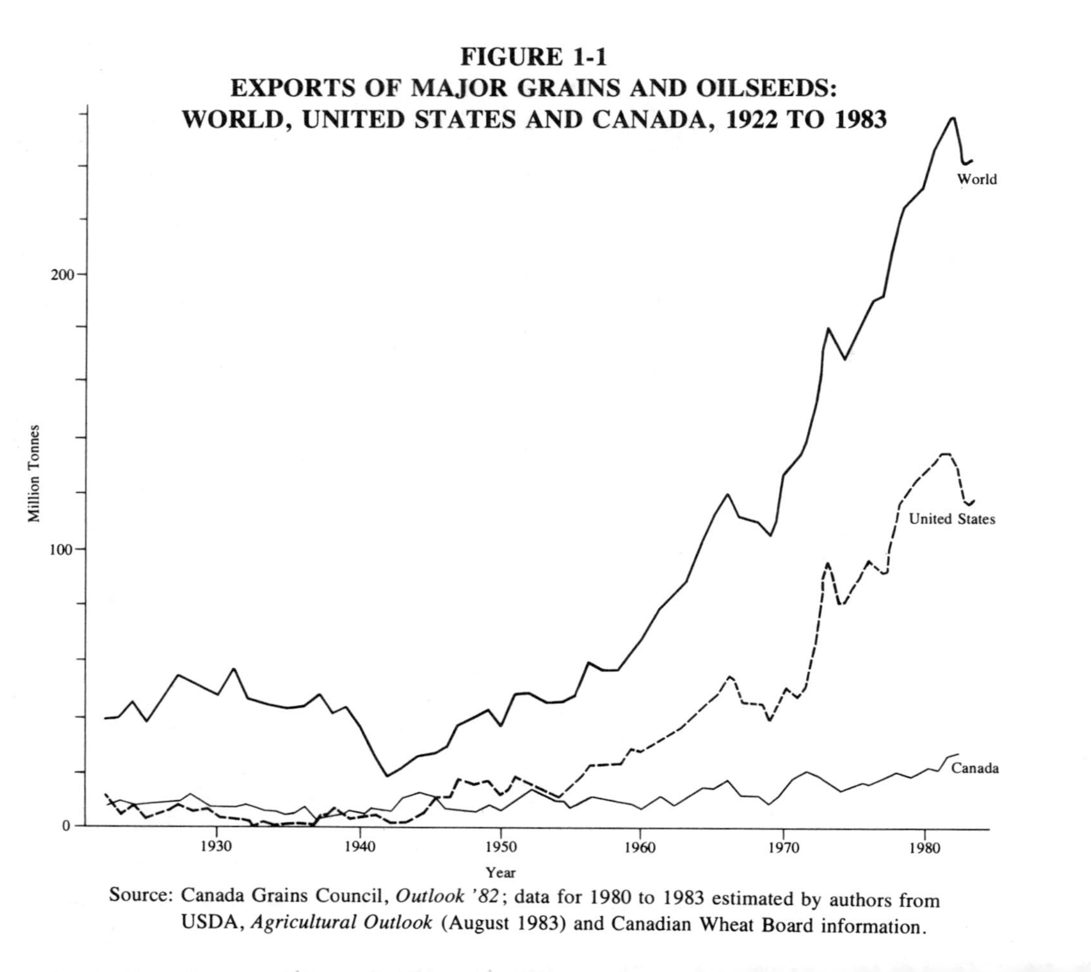

Source: Canada Grains Council, *Outlook '82*; data for 1980 to 1983 estimated by authors from USDA, *Agricultural Outlook* (August 1983) and Canadian Wheat Board information.

would seem that the Canadian Wheat Board targets are overly optimistic. However, during the same period, Canada achieved successive record levels of exports of prairie grain and oilseeds — 26.1 million tonnes in 1981-82 and an estimated 28 to 29 million tonnes in 1982-83. Such achievements, admittedly associated with years of good weather and production levels considerably above trend, are remarkably in step with the possibility of grain export levels in the 34 to 36 million tonnes range by 1990. What will the future hold? Further assessment is needed of the probable course of world trade in grains in the medium-run, the determinants of Canada's market share and of the production, transportation and policy difficulties that may constrain Canada's ability to achieve such grain export goals.

A central thrust of this study is an examination of the economic impacts that might be entailed in a moderate expansion in grain production and exports. Grains expansion may generate direct impacts on GNP, employment and trade; it may also induce economic activity in industries which are either forwardly linked or backwardly linked to the grains sector. The major forward-linked activities include the impacts on livestock feeding, the possibilities for further processing of grain prior to domestic use or export, and the scope for further oilseed crushing. The most important backward-linked industries are those supplying manufactured or off-farm inputs — agricultural machinery, pesticides and fertilizer. One question of interest here is whether more farm machinery and pesticides could be produced within the Canadian economy. Currently, some 80 percent of the farm machinery and more than 90 percent of the agricultural chemicals purchased in Canada are imported, primarily from the United States.

The capturing of more spin-off benefits from moderate grains expansion is likely to be a slow and difficult process. The prospects for increasing the degree of processing and the magnitude of value-added associated with Canadian grain exports are currently limited by the fact that 90 percent of exports of grain and grain products is in bulk or unprocessed form. The prospects of import substitution for farm inputs are limited by serious questions of cost, economic feasibility and economic efficiency.

Expanding prairie grain production will require many institutional, technical and policy innovations and reforms. Continued efforts to improve the grain handling and transportation system are necessary. (The issues involved in changes to the Crowsnest Pass rates, while not a major focus of this study, are briefly outlined.) Continued improvement in stabilization policy for grains will be needed to assist farmers in attaining greater income security and enhanced flexibility to meet production opportunities as they arise. The viability of grains expansion is currently undercut by sluggish Canadian and world economic growth, by the slow drift to increased protectionism in agricultural trade, and by less-than-adequate attention to research, to increasing productivity and to improvement of human skills relating to the grains sector. Finally, grains expansion will necessitate biochemical innovation at the farm level and raises various environmental concerns.

The future for the grains and oilseeds sector in Canada is moderately optimistic. The grains sector will not be the vitalizing force for the Canadian economy in the last two decades of this century that it was in the first three decades. Nevertheless, modest grains expansion presents growth opportunities which should not be neglected and which can be more fully realized if policy improvements for the grains sector are undertaken.

The Grains Economy: An Overview

2

A brief summary of the main features of the grains and oilseeds industry in Canada is presented in this chapter. The production of major grain and oilseed crops is described and briefly compared to production levels in other major grain producing nations. The contribution of grains and oilseeds to agricultural and overall trade is summarized, and major domestic uses in livestock feeding and further processing are introduced. Finally, the main aspects of the grain handling and transportation system in Canada are described, and a brief introduction to the nature and functions of some key grain institutions, including the Canadian Wheat Board and the Canadian Grain Commission, is provided.

The Producing Sector

Western Canada, specifically the prairies, dominates the production of grains and oilseeds in Canada. In 1981-82 the prairie provinces accounted for 96 percent of Canadian wheat production and 65 percent of total coarse grain production.[1] Most Canadian barley is produced in the prairies (92 percent in 1981-82) as is most of the oats (79 percent) and rye (89 percent). In contrast, corn is produced mainly in Ontario and Quebec; production of this crop has been increasing and by 1981-82 it accounted for 27 percent of coarse grain production in Canada. Rapeseed, the major oilseed crop, is virtually all produced in the prairie provinces as is much of the flaxseed, the second leading oilseed export. Soybeans, the second-ranking oilseed crop in terms of Canadian production, are grown only in Ontario.

The 1981 Census of Agriculture indicates that, of the total 318,361 Canadian farms, nearly half are located in the prairies (see Table 2-1). Prairie farms are relatively land-extensive with an average size of 244

TABLE 2-1
CANADIAN AND PRAIRIE AGRICULTURE, 1981

	Canada	*Prairies*
Total number of farms	318,361	154,816
Number of farms with sales of $2500 or more	271,604	142,023
Number of farms: wheat[1]	55,780	54,579
Number of farms: small grains[1]	52,086	35,188
Improved Area (million hectares)		
Under crops	31.0	24.6
Summer fallow	9.7	9.5
Pasture	4.4	2.9
Total[2]	46.1	37.7
Improved area per farm (hectares)	144.9	243.6
Cropped Area (million hectares)		
Wheat	12.4	12.1
Barley	5.5	5.0
Other grains[3]	3.7	1.9
Total grains	21.6	19.1
Rapeseed	1.4	1.4
Other oilseeds[4]	1.0	0.7
Total Oilseeds	2.4	2.1

Notes: [1] Farms with sales of $2,500 or more, by major product sold.
[2] Including other improved land uses.
[3] Oats, mixed grains, corn, rye, buckwheat.
[4] Flaxseed, soybeans, sunflowers, mustard seed.
Source: Statistics Canada, *1981 Census of Canada*.

improved hectares, which is larger than the average Canadian farm size of 145 improved hectares. Nearly all prairie farms are family farm operations in that most farms, particularly commercial farms, are operated by their owners, frequently with the assistance of other family members and sometimes with hired workers. Slightly more than one-third of prairie farms include some rented land, and these tend to be larger than the average. For many farmers, particularly those with smaller farms, off-farm employment is a major source of income. Prairie farms have gradually increased in size while decreasing in numbers. As is the case for all Canadian farms, over the 1970s they showed substantial increases in capital values whether expressed in nominal or real value terms. Both the substitution of capital for labour and the process of technical change has, particularly since the 1940s, been associated with farm mechanization and with the increasing size and technical sophistication of farm machinery.

As seen in Table 2-1, 40 percent of Canada's 271,604 farms with gross sales of $2,500 or more can be considered grain and oilseed farms — nearly 56,000 producing wheat (virtually all located on the prairies) and 52,000 growing other grains and oilseeds (some 68 percent of these are on the prairies). In 1981 the typical grain and oilseed farm in Canada had 268 hectares of improved land and an average capital value of $532,000, 83 percent of which was the value of land and buildings.[2]

Data for 1960 to 1982 on the area and production of the two most significant grain crops — wheat and barley — and of the most important oilseed crop — rapeseed or canola — are presented in Table 2-2. The current dominance of wheat, which accounts for approximately one-half of the area sown to grains and oilseeds in Canada, is clearly evident. Wheat acreage dropped sharply in 1970 due to Operation LIFT (Lower Inventories for Tomorrow), a one-year government program to hold land out of production, and only recovered to the levels of the 1960s in 1976. About 85 percent of Canada's wheat acreage is planted to hard red spring wheat varieties, with approximately 12 percent sown to durum wheat, 2 percent to winter wheat, and a very small but increasing fraction to soft white spring wheat. Production of all wheats reached record highs in 1981 and 1982 — nearly 25 million tonnes in 1981, followed by nearly 27 million tonnes in 1982. Canadian wheat yields, approaching two tonnes per hectare, are marginally higher than the world average but are only 85 percent of average yields in the United States and 40 percent of yield levels in France.[3]

Barley plantings have averaged nearly 5 million hectares annually in the past decade, while annual barley production has averaged nearly 11 million tonnes. Canada has gradually increased its share of the world non-corn coarse grain market. The "Cinderella story" associated with rapeseed — improved varieties are now called canola — is seen in the sharp increases in acreage and production which began in 1970. While the sown area and production levels of rapeseed have dropped in the early 1980s, the relative area sown to oilseeds on the prairies is expected to increase by 1990.

TABLE 2-2
AREA AND PRODUCTION OF WHEAT, BARLEY AND RAPESEED, CANADA, 1960 TO 1982

	Area (million hectares)			*Production (million tonnes)*		
	Wheat	*Barley*	*Rapeseed /Canola*	*Wheat*	*Barley*	*Rapeseed /Canola*
1960	9.9	2.8	0.3	14.1	4.2	0.2
1961	10.2	2.2	0.3	7.7	2.4	0.2
1962	10.8	2.1	0.1	15.4	3.6	0.1
1963	11.2	2.5	0.2	19.7	4.8	0.2
1964	12.0	2.2	0.3	16.3	3.7	0.3
1965	11.4	2.5	0.6	17.7	4.7	0.5
1966	12.0	3.0	0.6	22.5	6.4	0.6
1967	12.2	3.3	0.7	16.1	5.5	0.6
1968	11.9	3.6	0.4	17.7	7.1	0.4
1969	10.1	3.8	0.8	18.3	8.1	0.8
1970	5.0	4.0	1.6	9.0	8.9	1.6
1971	7.8	5.7	2.1	14.4	13.1	2.1
1972	8.6	5.1	1.3	14.5	11.3	1.3
1973	9.6	4.8	1.3	16.2	10.2	1.2
1974	8.9	4.8	1.3	13.3	8.8	1.2
1975	9.5	4.5	1.8	17.1	9.5	1.8
1976	11.3	4.3	0.7	23.6	10.5	0.8
1977	10.1	4.7	1.4	19.9	11.8	2.0
1978	10.6	4.3	2.8	21.1	10.4	3.5
1979	10.5	3.7	3.4	17.2	8.5	3.4
1980	11.1	4.6	2.1	19.2	11.3	2.5
1981	12.4	5.5	1.4	24.8	13.7	1.8
1982[1]	12.6	5.2	1.7	26.9	13.6	2.1

Note: [1] Preliminary.

Source: Statistics Canada, *Handbook of Agricultural Statistics: Field Crops, 1921-1979*; and Canada Grains Council, *Canadian Grains Industry Statistical Handbook 1982*.

Canada: A World Bread-basket?

Many Canadians have the image of Canada as a leading world producer of grains. In part, this image arises from wheat's important place in the international trade in grains and Canada's considerable role in providing 20 percent of world wheat exports. Canada, however, is more important as a trader of grains than as a world producer.

The three most important grains in the world are wheat, corn and rice — with 448, 446 and 404 million tonnes of each being produced, respectively, in 1982.[4] The leading producer of wheat, which is surprising to many, is the USSR with 80 million tonnes in 1982, well below its previous ten-year average of 93 million tonnes. In second place in 1982 was the United States with an estimated 75 million tonnes, some 20 million tonnes above its previous ten-year average. In 1982 China produced 58.5 million tonnes of wheat, while India, tripling its wheat output during the past two decades due to the high-yielding varieties associated with the green revolution, grew 36.5 million tonnes. The record level of wheat output of 26.9 million tonnes put Canada in fifth place and comprised only 6 percent of world wheat production.

While Canada is the second largest world producer of barley (considerably behind the USSR), the coarse grain market is dominated by corn, nearly half of which is produced in the United States. Similarly, while Canada, along with China and India, are the major producers of rapeseed, the international oilseed market is dominated by soybeans — again mostly produced in the United States. The great bulk of world rice production, little of which is traded internationally, occurs in Asia, with China generating 144 million tonnes in 1982. Nor is Canada involved to any extent in sorghum and millet production, two "inferior" cereals which are significant in the diets of the poor in drier regions of the Third World. Overall, Canada produces only about 3 percent of the world's grain.

The Export and Domestic Use of Grains

Grains and oilseeds produced in Canada are either exported to overseas markets or used domestically. The leading Canadian grain and oilseed exports are wheat, barley and rapeseed (see Figure 2-1). In 1982-83 the Canadian Grain Commission estimated that Canadian bulk exports of grains and oilseeds (excluding corn and soybeans, and also excluding wheat flour which is typically included in Canadian Wheat Board estimates) totalled a record 28.3 million tonnes. Of this total, bulk wheat exports (including durum wheat) constituted 21 million tonnes or nearly 75 percent — a somewhat higher share than has been typical.

The major customers for Canadian grains and oilseeds for three recent crop years (1980-81 through 1982-83) are portrayed in the bar graphs in Figure 2-2. The USSR is currently Canada's leading grain export market, followed by China and Japan. Other major buyers of

FIGURE 2-1
CANADIAN GRAIN AND OILSEED EXPORTS, 1982-83

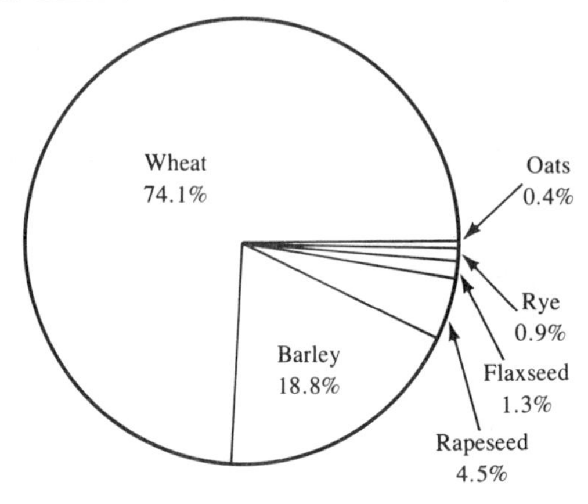

Source: Based on preliminary estimates (excluding corn and soybeans) of the Canadian Grain Commission as reported in *Globe and Mail*, 11 August 1983

Canadian grains and oilseeds are Brazil, Cuba, the United Kingdom, Italy and Poland. The most notable trends in Canada's grain trade in the early 1980s are the emergence of the Soviet Union as Canada's dominant market, the continued growth in sales to China, and the continued slippage of the United Kingdom as a major market.

In recent years, the export of agricultural products has amounted to 10 to 11 percent of the value of Canadian merchandise exports (see Table 2-3). These agricultural exports are dominated by grains and oilseeds, most of which are exported in non-processed forms. Grains, oilseeds and their products have contributed approximately 70 percent of the value of all agricultural exports from Canada in recent years. Export of grains and grain products alone accounted for nearly 60 percent of all agricultural exports in 1981 and 1982. Less than 10 percent of grain and grain product exports was in the form of processed products — mainly flour and animal feeds.

Exports of oilseeds and oilseed products have accounted for somewhat more than 10 percent of all agricultural exports in recent years (except in 1982 when the share decreased to 8.6 percent). Exports are mainly of unprocessed seed (mainly rapeseed, followed by flaxseed), although the proportion of rapeseed exported in processed or semi-processed forms has been increasing — these accounted for an average of 11 percent of the value of oilseed and oilseed product

FIGURE 2-2
MAJOR IMPORTERS OF CANADIAN GRAIN, 1980-81, 1981-82, 1982-83

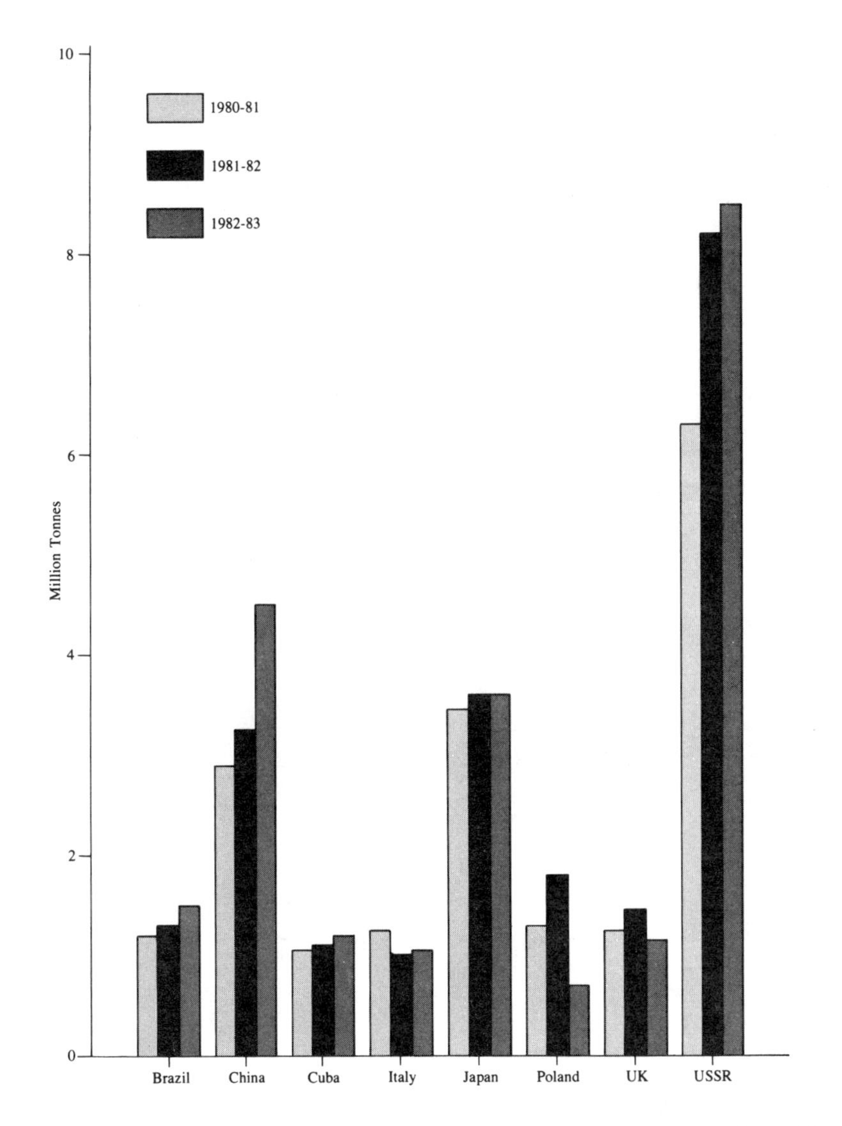

Source: Canadian Grain Commission estimates as reported in *The Western Producer*, 22 September 1983, p.4.

TABLE 2-3

CANADIAN AGRICULTURAL TRADE: EXPORTS AND IMPORTS BY MAJOR AGRICULTURAL COMMODITY GROUP (millions of current dollars)

	Exports				*Imports*			
	1971-75 (average)	*1976-80 (average)*	*1981*	*1982*[1]	*1971-75 (average)*	*1976-80 (average)*	*1981*	*1982*[1]
Grains and grain products[2]	1,859	3,061	5,201	5,577	13	213	374	290
(Wheat and wheat flour)	(1,411)	(2,326)	(3,723)	N/A	—	—	—	—
Oilseeds and oilseed products	343	677	1,004	804	222	411	506	439
Animal feeds	71	142	199	214	25	44	64	78
Animals, meat & other animal products	370	843	1,211	1,479	364	638	811	680
Dairy products	64	106	209	285	52	74	94	101
Poultry and eggs	16	27	53	42	21	60	70	75
Fruits, nuts and vegetables	95	200	384	402	599	1,230	1,811	1,882
Other[3]	178	321	522	503	729	1,431	1,884	1,513
Agricultural Total	2,996	5,377	8,783	9,306	2,145	4,101	5,614	5,058
All Commodities	25,246	54,472	81,337	81,829	24,865	52,422	79,482	67,926
Share of grains and grain products in total agricultural exports/imports (%)	62.0	56.9	59.2	59.9	6.2	5.2	6.7	5.7
Share of oilseeds and oilseed products in total agricultural exports/imports (%)	11.4	12.6	11.4	8.6	10.3	10.0	9.0	8.7
Share of agricultural products in total commodities exports/imports (%)	11.9	9.9	10.8	11.4	8.6	7.8	7.1	7.4

Notes: [1] Preliminary. [2] The conventional agricultural trade statistics do not appear to include malted barley.
[3] Includes seeds for sowing, maple products, honey, sugar, tobacco, vegetable fibres, plantation crops and all other agricultural products.

Sources: Agriculture Canada, *Canada's Trade in Agricultural Products* (various years); Agriculture Canada, *Selected Agricultural Statistics, Canada and the Provinces, 1983*; Statistics Canada, *Exports Merchandise Trade*; and Statistics Canada, *Imports Merchandise Trade*.

exports from 1971 to 1975 but reached 20 percent of the value of Canadian oilseed and product exports in 1981. Rapeseed that is not exported as raw seed is crushed to yield some 42 percent, by weight, of oil which is generally further refined into a variety of vegetable oil products, such as margarine, salad dressing, cooking oils and shortening. The residue from crushing forms a high protein meal (40 percent protein content) used for livestock feed. Canada also imports oilseed products and raw oilseed — mainly soybeans — from the United States and the value of these imports is approximately half that of Canadian oilseed and product exports. The proportion of processed oilseed products is higher for imports than for exports but the proportion of processed imports has declined slightly over the 1970s as the domestic crushing and processing industry has developed.[5]

While export demand constitutes the most substantial markets for the grains sector, domestic use, particularly by the livestock sectors, involves significant quantities of Canadian grains. The disposition of Canadian wheat and barley production in 1981-82 to export and to various domestic uses is portrayed in Figures 2-3 and 2-4, respectively. Expressed as a percentage of volume produced, some 25 to 35 percent of wheat production is consumed domestically and nearly two-thirds of this is used in the domestic livestock industry. For the feed grains of

FIGURE 2-3
DISPOSITION OF CANADIAN WHEAT PRODUCTION, 1981-82

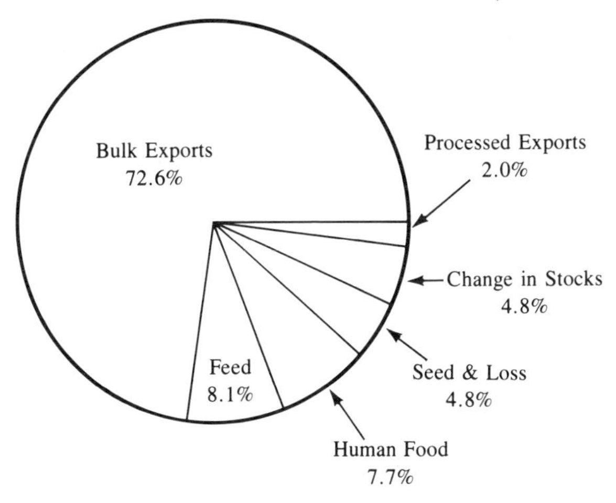

Source: Tables 6-1, 6-2 and 6-3

FIGURE 2-4
DISPOSITION OF CANADIAN BARLEY PRODUCTION, 1981-82

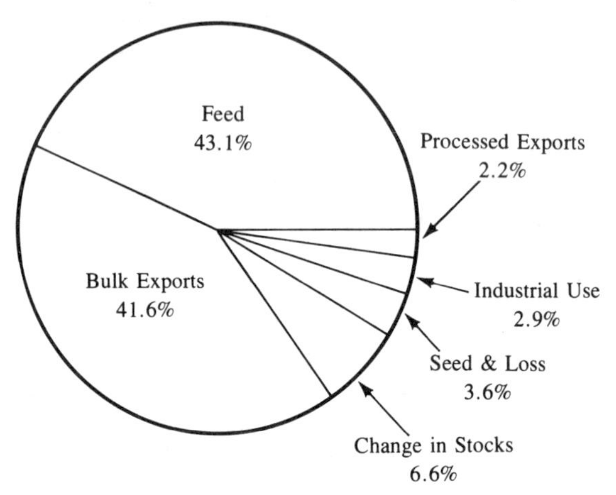

Source: Tables 6-1, 6-2 and 6-3

barley and oats, domestic use as a livestock feed constitutes the largest market, accounting for about 60 to 65 percent of the barley and, on average, about 90 percent of the oats produced in Canada over the 1970s. For rapeseed, the major oilseed produced in Canada, domestic use increased substantially over the late 1970s and by the early 1980s this has approached the level of exports of this crop.

Prairie grains are a major intermediate input — whether produced on-farm or purchased — to some 39,000 prairie livestock farms as well as being a significant purchased input for many of the 80,000 other Canadian livestock farms. Other forward-linked impacts of the western grains economy occur through the relatively small domestically oriented secondary processing of grains and oilseeds for human food and industrial uses. Impacts on the Canadian economy from this sector also arise from the handling and distribution system for grains and oilseeds. This involves elevators, railways, ports and their associated operating and regulatory institutions. The backward-linked impacts of the western grains sector primarily arise from the increasing use of purchased off-farm nonfeed inputs associated with modern agriculture, the most important of which are machinery, fertilizers and other farm chemicals. The nature and strength of these linkages of the grains and oilseeds sector to the Canadian economy are considered in more detail in chapters 5, 6, and 7.

The Grain Handling and Transportation System

In 1981-82 the grain handling system took delivery of some 31 million tonnes of the major grains and oilseeds from some 146,000 holders of Canadian Wheat Board permit books.[6] Most of these deliveries were to the 2,934 licensed primary elevators at 1,217 elevator points scattered throughout the grain growing areas of the prairies on some 25,000 kilometres of track.[7] Primary elevators purchase grain from producers, either on behalf of the Canadian Wheat Board or the company itself, weigh, sample, grade, elevate and load this grain. They may clean and store grain before forward delivery through the rail system to terminals at Thunder Bay, Vancouver, Prince Rupert and Churchill or for domestic processing or use. More than three-quarters of the licensed primary elevators are owned and operated by the four farmer-controlled cooperative grain handling companies. Saskatchewan Wheat Pool owns 32 percent of licensed primary elevators (these account for 28 percent of the total storage capacity of 8,138 thousand tonnes of licensed primary elevators). Alberta Wheat Pool owns 21 percent, while Manitoba Pool Elevators owns 7 percent and United Grain Growers owns 17 percent of primary elevators. The relatively small private grain handling companies of Pioneer Grain, N.M. Paterson & Sons, and Parrish & Heimbecker account for, respectively, 12 percent, 3 percent and 2 percent of primary elevators, while Cargill Grain, a large multinational private grain-trading company, comprises 6 percent.

The number of times that an elevator handles its capacity during a crop year (its turnover ratio) generally ranges from three to four; turnover rates are expected to gradually increase with continued consolidation of elevators. The number of primary elevators has declined — there were 5,145 in 1965 compared to 2,934 in 1982 — as numbers of outdated, low-volume and uneconomic elevators and rail branchlines have gradually closed.

There are 24 licensed terminal elevators with a capacity of 3,695 thousand tonnes. Most of these are located at export points, although 5 inland elevators are located on the prairies. The terminal elevators grade, clean, dry and accumulate grains in sufficient quantities for export lots. The elevator network for western grains also includes 27 transfer elevators in Eastern Canada (with 3,690 thousand tonnes capacity). As well as providing facilities for handling western grains, they handle eastern grain and grain from the United States. There are also twenty-four licensed process elevators which receive and store grain and oilseeds for processing into such products as flour, animal feed, malt, or crude vegetable oil and meal. In addition, there were forty licensed grain dealers in 1982. The various elevators and grain dealers for western grain are licensed by the Canadian Grain Commission which, among other duties, specifies the maximum tariffs that licensed elevators can charge.

Trucks, generally owned and operated by farmers, are used to move grains and oilseeds to primary elevators. Virtually all grains are moved over larger distances through the rail system, primarily by Canadian National Railways, Canadian Pacific Railway, and British Columbia Railway. Regulated rates — the longstanding Crowsnest Pass rates — have applied on rail shipments of designated grains and products to specified export points (this issue is discussed in chapters 6 and 8). Grains moved eastward through the Great Lakes and St. Lawrence Seaway system are shipped in laker vessels. About ninety lakers carry grain from Thunder Bay to other ports on the Great Lakes and St. Lawrence system.

More than half of the export grain moves eastward through Thunder Bay, the country's largest grain handling port, and is shipped for export from St. Lawrence ports. Smaller quantities are loaded on ocean vessels for direct export from the Lakehead. Exports from Pacific ports, particularly from Vancouver which is the second largest grain handling port, are close rivals to those from the St. Lawrence. Export shipments through Churchill face major geographic constraints (this port is generally only open for about ninety days each year) as do shipments from the Atlantic seaboard ports. To encourage continued use of the Atlantic ports for export of grain and flour, rail tariffs on

TABLE 2-4
GRAIN AND OILSEED EXPORTS BY PORT, 1981-82 AND 1982-83
(million tonnes)

	1981-82	*1982-83*[1]
Pacific Seaboard	11.63	11.52
Thunder Bay	0.91	0.61
St. Lawrence	11.95	14.62
Atlantic Seaboard	0.86	0.66
Churchill	0.44	0.56
U.S. shipments	0.26	0.34
Total	26.05	28.30

Note: [1] Preliminary.
Source: *Globe and Mail*, 11 August 1983, based on Canadian Grain Commission estimates.

shipments of grain and flour from inland positions to these ports are subsidized. Rail carriers charge freight rates that applied in 1960 for grain and rates that applied in 1966 for flour. The differences between these and compensatory freight rates are determined and paid by the Canadian Transport Commission. The expenditures for this subsidy have been increasing; they amounted to $28 million in 1980.[8] Shipments from the various ports are shown in Table 2-4.

The Canadian Wheat Board

The Canadian Wheat Board (CWB), established in 1935, is a crown corporation directed by appointed commissioners, who are advised by a committee of elected producers. It represents the interests of prairie grain producers and has, throughout its history, been a major agent of government policy for western grains.

Forerunners of the board were established on a temporary basis during the disrupted market conditions which prevailed during and immediately after the First World War. The 1935 board was a voluntary marketing agency for prairie wheat; it did not acquire compulsory marketing powers until 1943, initially under the War Measures Act, as a means of enforcing wartime price controls. The board's powers were extended to oats and barley in 1949.

The CWB is the sole purchaser and marketing agency for export marketings of prairie wheat, oats and barley and for domestic human uses of these crops. Its powers relate to the three specified grains which are produced in the designated area of Manitoba, Saskatchewan, Alberta and the Peace River area of British Columbia. It does not generally own or operate physical marketing facilities but uses, as agents, the various grain handling companies who buy, handle and frequently sell grain on behalf of the CWB. The CWB administers the system of initial payments to grain producers (these are established and guaranteed by the federal government). It operates annual price averaging systems, or pools, so that producers receive an average price which is adjusted for grade.

The board administers a system of delivery quotas; these quotas essentially relate farmers' delivery opportunities to farm size (in terms of cultivated area) and regulate farmers' deliveries of the major prairie grains and oilseeds to the elevator system. This system was originally intended to ration delivery opportunities fairly among farmers. Delivery quotas also serve as a means for the CWB to bring producers' deliveries in line with transport availability and sales opportunities. The quota system has been criticized for being unduely restrictive at times, and for tending to discourage productivity growth in periods when quotas have been restrictive. While delivery quotas may well be a factor affecting subsequent planting decisions by producers, they relate to deliveries and are not entry-limiting or directly supply-restricting as are the marketing quotas applied by the supply management marketing boards for dairy and poultry products.

Since the establishment of the Grain Transportation Authority in 1979, that agency and its coordinator have assumed overall responsibility for allocation of the grain car fleet. The CWB administers the block shipping system for the movement of wheat and the other major grains and oilseeds from primary elevator positions to terminal elevators or mills.[9]

The CWB sells grain for export through accredited exporters (including the major Canadian grain handling companies and international grain trading companies); it also sells grain directly, generally to state trading companies, and often under long-term sales agreements. Its direct sales now account for a high proportion of Canadian grain exports (particularly for wheat and barley), which is a reversal of the situation that prevailed during the 1950s and 1960s.

The Canadian Grain Commission

The Canadian Grain Commission has regulatory authority over the quality of Canadian grain and over its handling, transportation and storage system. Its main activities involve the inspection and grading of grain and the licensing of elevators and other grain trade agencies in Canada. The commission also specifies the maximum tariffs for the handling, storage and cleaning of grains. This body also regulates the allocation of producer-loaded grain cars.[10]

The official inspection of grain is visually based and the development and licensing of new varieties places much emphasis on their visual distinguishability from other varieties. This system has promoted a remarkable homogeneity of Canadian grain types, particularly of wheat, so that a small number of varieties account for the majority of production. Leading varieties are shown in Figure 8-2. The regulatory institutions for the marketing and quality control of Western Canadian grain have placed major emphasis on maintenance of Canada's reputation for producing and selling high quality bread wheats. Varietal development has not emphasized a search for and adaptation of higher yielding grains suited to Canadian conditions.

International Markets for Canadian Grains

3

The export sector for grains and oilseeds is of fundamental importance to the Western Canadian grains economy. Some 77 percent of wheat produced in Canada is exported, as is 36 percent of Canadian barley production and 59 percent of rapeseed/canola.[1] Changes in export demand for grains have an immediate and direct effect on the revenues and hence on the economic fortunes of grain growers; they also have an immediate though indirect effect on the input costs of most livestock producers and on other forward- and backward-linked sectors. Export demand for grains and oilseeds has generally grown and projections indicate further growth. The nature of this export demand and future prospects in export markets are thus of major importance to Canadian grain producers and the sectors with which they are linked.

The Changing Structure of International Markets

Prior to decreases in world grain trade in 1982-83, world trade in wheat and coarse grains expanded very rapidly. Over the two decades prior to 1982-83, the volume of grain entering world markets more than trebled (see Tables 3-1 and 3-2). Increasing levels of population and increasing affluence contributed to the increased trade in grains during this period. The volumes of grain traded increased more rapidly than either world production or world consumption. The increases in trade and in production and consumption were greatest for coarse grains, particularly corn, a feature that reflects the preference of consumers to improve their diets by substituting animal for vegetable protein as incomes increase — much of the traded coarse grains is used for animal feed. In 1981 corn accounted for about 68 percent of world coarse grain exports while barley accounted for 17 percent.[2]

Wheat is the food grain that is most frequently traded in international markets;[3] world trade in wheat grew at about 6 percent per year over the 1970s.[4] Wheat is not a homogeneous commodity. It differs in

TABLE 3-1: WORLD WHEAT SUPPLY/DEMAND, 1960-61 TO 1982-83 (million tonnes/hectares)

	Area Harvested	*Yield*	*Production*	*July/June Trade*[1]	*Estimated Total Utilization*	*Estimated Ending Stocks*	*Stocks as % of Utilization*	*Trade as % of Production*
1960-61	202.2	1.18	238.4	41.9	234.8	81.8	34.8	18
1961-62	203.4	1.10	224.8	46.8	236.3	70.2	29.7	21
1962-63	206.9	1.22	251.8	44.3	248.1	74.0	29.8	18
1963-64	206.3	1.13	233.9	56.0	240.0	67.8	28.3	24
1964-65	215.9	1.25	270.4	52.0	262.0	76.2	29.1	19
1965-66	215.5	1.22	263.3	61.0	281.5	55.3	19.7	23
1966-67	213.7	1.44	306.8	56.0	279.9	82.1	29.4	18
1967-68	219.3	1.36	297.6	51.0	289.1	90.6	31.3	17
1968-69	223.9	1.48	330.9	45.0	306.5	115.0	37.6	14
1969-70	217.8	1.42	310.0	50.0	327.2	97.8	30.0	16
1970-71	207.0	1.52	313.8	55.0	337.3	74.3	22.0	18
1971-72	212.9	1.65	350.9	52.0	344.2	81.0	23.5	15
1972-73	211.2	1.63	343.5	67.0	361.8	62.6	17.3	20
1973-74	217.0	1.72	373.0	63.0	365.4	70.2	19.2	17
1974-75	220.1	1.64	360.2	64.3	366.4	64.0	17.4	18
1975-76	225.4	1.58	356.5	66.7	356.2	64.1	18.0	19
1976-77	233.2	1.81	421.3	63.3	385.8	99.8	25.9	15
1977-78	227.1	1.69	384.1	72.8	399.3	84.3	21.1	19
1978-79	228.9	1.95	446.8	72.0	430.2	100.9	23.4	16
1979-80	227.6	1.86	422.8	86.0	443.5	80.4	18.1	20
1980-81	236.6	1.86	441.1	94.1	446.5	78.7	17.6	21
1981-82[2]	239.3	1.88	448.9	101.7	442.1	85.5	19.3	23
1982-83[3]	238.8	2.01	480.3	97.7	468.3	97.4	20.8	20
1983-84[3]	—	—	478.4	98.6	446.8	109.0	23.3	21

Notes: [1] Trade data exclude intra-EEC trade. [2] Preliminary. [3] Projection.

Source: USDA, *Foreign Agricultural Circular* (March 1983), p. 24; and USDA, ERS, *Agricultural Outlook* (August 1983), p. 47.

TABLE 3-2: WORLD COARSE GRAINS SUPPLY/DEMAND, 1960-61 TO 1982-83 (million tonnes/hectares)

	Area Harvested	*Yield*	*Production*	*July/June Trade*[1]	*Estimated Total Utilization*	*Estimated Ending Stocks*	*Stocks as % of Utilization*	*Trade as % of Production*
1960-61	324.4	1.38	447.9	24.0	437.2	109.7	25.1	5
1961-62	322.4	1.35	434.2	30.0	449.3	94.7	21.1	7
1962-63	320.9	1.43	459.5	31.0	461.5	92.7	20.1	7
1963-64	326.5	1.43	467.7	34.0	462.5	97.9	21.2	7
1964-65	323.5	1.46	472.6	35.0	479.5	90.9	19.0	7
1965-66	320.1	1.51	484.7	42.0	500.5	75.1	15.0	9
1966-67	321.9	1.62	521.2	40.0	520.2	76.1	14.6	8
1967-68	327.3	1.68	551.4	39.0	542.3	85.2	15.7	7
1968-69	326.8	1.69	552.6	37.0	548.6	89.2	16.2	7
1969-70	330.7	1.74	576.7	39.0	576.6	89.2	15.5	7
1970-71	331.8	1.74	576.3	46.0	593.3	72.2	12.2	8
1971-72	333.4	1.89	629.1	49.0	615.4	87.0	14.2	8
1972-73	329.1	1.85	609.9	59.0	626.9	69.9	11.1	10
1973-74	344.5	1.94	669.7	71.0	673.0	65.9	9.8	11
1974-75	342.1	1.84	628.1	64.9	633.6	58.9	9.3	10
1975-76	348.3	1.85	645.0	75.1	645.6	58.3	9.0	12
1976-77	343.7	2.05	704.2	82.7	685.3	77.5	11.3	12
1977-78	345.1	2.03	700.6	84.0	692.0	85.9	12.4	12
1978-79	342.8	2.20	753.6	90.2	748.1	91.2	12.2	12
1979-80	341.1	2.17	741.5	100.9	740.3	91.6	12.4	14
1980-81	342.3	2.13	730.0	105.5	740.8	80.9	10.9	14
1981-82[2]	349.1	2.19	764.8	105.4	732.3	113.3	15.5	14
1982-83[3]	341.7	2.28	780.9	88.5	784.4	145.8	18.6	11
1983-84[3]	—	—	743.0	94.2	779.5	109.4	14.0	13

Note: [1] Trade data exclude intra-EEC trade. [2] Preliminary. [3] Projection.

Source: USDA, *Foreign Agricultural Circular*, p. 24; and USDA, *Agricultural Outlook* (1983), p. 47.

protein content and milling characteristics, and thus in use, as well as in its agronomic characteristics, and thus the areas where different varieties are grown. Approximately two-thirds of world wheat is used for human food; most of the rest is used for animal feed.

Over the 1970s there was a tendency for the level of world grain stocks, relative to utilization, to decline (see Tables 3-1 and 3-2). During the 1960s, grain prices were relatively depressed while substantial stocks of grain were accumulated in the major exporting nations. Policy initiatives by governments and their agencies in major grain exporting nations to reduce the level of expensive stock-holding of grains were instituted in the late 1960s and early 1970s. Coincidentally, a series of poor crops in some major grain producing and grain consuming nations occurred in the early 1970s. At the same time internal policy changes led the USSR to rely increasingly on imports to offset highly variable grain deficits. All these factors contributed to the decline in world grain stocks and to the major increases in grain prices on international markets in 1973 and 1974 (see Figure 3-1).

World market prices for cereal grains fell from 1975 to 1977 as world grain production and stock levels increased. Subsequently, until early 1981 grain prices increased, reflecting increased volumes of trade and decreased stocks, features that were influenced by the decreased production and increased importation of grains by the USSR from 1979 to 1981. Grain prices again fell through 1982, reflecting demand dampened by world recession as stocks increased, particularly in the United States, and these tendencies have continued into 1983. Closing stocks for 1981-82 reached record levels, although ending season stock-utilization ratios (15 percent for 1981-82 and 20 percent anticipated for 1982-83)[5] are somewhat less than those which applied through most of the 1960s. (The Food and Agriculture Organization (FAO) has suggested that a stock-utilization ratio of 17 to 18 percent provides short-term global food security.) Overall, a high level of variability of world grain prices has applied from 1973 to 1983. The nature of world grain markets is such that variability in grain prices is likely to continue to be a feature of world grain markets throughout the 1980s.

Changes in the pattern of trade flows in cereal grains are illustrated in Table 3-3. Over the 1970s, the United States became the world's dominant cereal exporting region. Third World and centrally planned countries have become the major cereal importers. In particular, the USSR has become the world's largest cereal importer although the

FIGURE 3-1
ACTUAL AND DEFLATED REAL AVERAGE PRICE RECEIVED FOR WHEAT PER TONNE: PRAIRIE PROVINCES, 1930-31 TO 1981-82

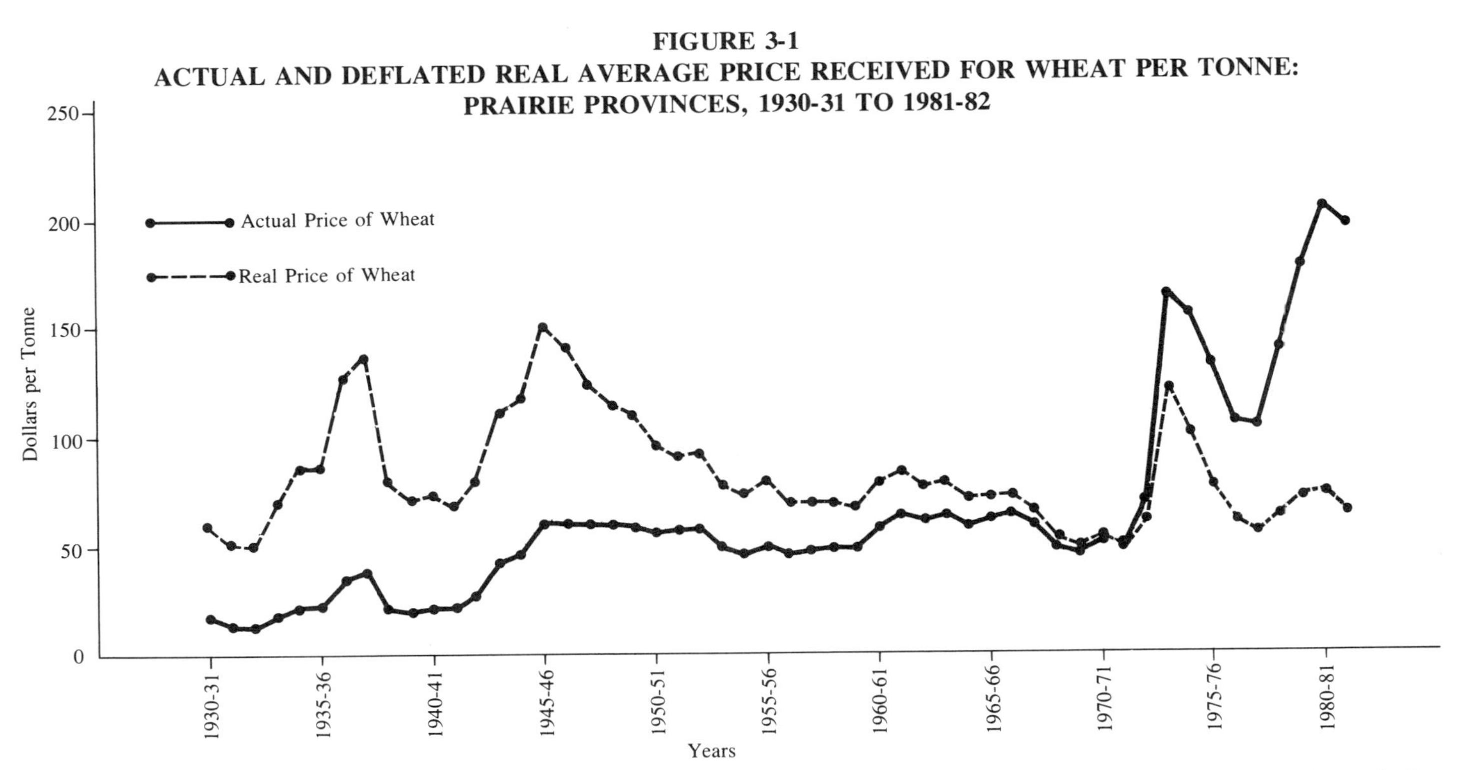

Note: The price deflator which is used is the Farm Input Price Index for Western Canada. The real price series, therefore, represents the relative purchasing power over time of a tonne of wheat in terms of inputs, not consumer goods.

Source: Based on data for 1930-31 to 1980-81 from Statistics Canada which is presented in J. C. Gilson, *Western Grain Transportation: Report on Consultations and Recommendations*, p. IV-36; 1981-82 data estimated by the authors.

TABLE 3-3
CHANGES IN NET TRADE IN GRAINS BY MAJOR TRADING REGIONS, SELECTED PERIODS FROM 1934-35 TO 1981
(million tonnes)

Region	*1934-35 to 1938-39 (annual average)*	*1951-52 to 1955-56 (annual average)*	*1956-57 to 1960-61 (annual average)*	*1961-62 to 1965-66 (annual average)*	*1970*	*1975*	*1980*	*1981*
North America	5.7	23.3	32.2	50.6	53.56	92.83	129.97	131.56
U.S.	0.8	12.0	22.5	38.0	38.24	78.13	109.65	110.08
Canada	4.9	11.3	9.7	12.6	14.42	14.70	20.13	21.28
Western Europe	−22.9	−21.3	−23.5	−26.2	−21.64	−19.14	−10.21	−3.45
France	−1.0	−0.5	−1.3	−4.6	9.31	12.08	18.30	20.62
Eastern Europe and USSR	4.9	0.9	0.4	−5.4	−1.38	−19.57	−41.49	−51.02
Central America	−0.4	−1.1	−1.3	−1.0	−2.70	−6.62	−11.02	−10.61
South America	9.6	2.1	2.4	4.0	6.91	3.13	−3.58	6.68
Asia, except China	0.8	−6.0	−11.4	−18.0	−29.14	−40.47	−47.28	−49.37
China	−0.3	0.1	−0.5	−5.7	−6.42	−5.61	−17.36	−17.05
Africa	1.0	0.0	−0.9	−2.5	−3.72	−8.77	−15.40	−17.23
Oceania	2.9	3.0	3.6	6.7	8.14	10.80	18.88	12.81

Note: A negative figure indicates net imports. Excludes rice.

Sources: F.A.O., *World Grain Trade Statistics*; and F.A.O., *Trade Yearbook*, various years.

levels of these imports are extremely variable. The protectionist common agricultural policy introduced by the European Economic Community (EEC) in 1967 has contributed to the overall tendency for Western Europe to become relatively more self-sufficient in cereal grains. This tendency has also been supported by changes in milling technology, which have made it possible for bakers to substitute more soft wheats for the hard red spring varieties. The EEC continues to be a substantial importer of coarse grains, particularly of corn, and has become a consistent net exporter of wheat and flour; hard wheat continues to be imported and increasing amounts of soft wheat are exported. France, in particular, has become a major wheat exporter, accounting for 10 to 13 percent of world exports of wheat and flour in recent years.

World Trade in Wheat

Canada is a major wheat exporter, although Canada's market share of world wheat exports has tended to fluctuate. Data from the U.S. Department of Agriculture on absolute levels of wheat exports for major exporters, as well as respective relative shares of world wheat exports, are presented in Table 3-4 and Figure 3-2. World trade in wheat is concentrated in the hands of five exporting nations: United States, Canada, France, Australia and Argentina. Together the United States and Canada provide 60 to 65 percent of total world wheat exports.

Canada's share of world wheat exports averaged about 22 percent during the 1960s, dropping somewhat toward the end of the decade. While some recovery occurred in the early 1970s, Canada's market share began to decline again beginning in 1973. At the same time the United States, the consistently largest exporter of wheat, increased its market share from approximately 40 to about 45 percent. Canada's declining, or at least below normal, market share throughout the 1970s is likely attributable to below-average production and availability of wheat in the 1970 to 1975 period and, even more importantly, to severe transportation bottlenecks from the mid-1970s to 1981. Starting in 1982-83, however, Canada's market share has rebounded to 21 percent and the share of the United States has fallen back close to its historic 40 percent level. Canada's recently improved wheat market performance is chiefly attributable to record-level production, absence of rail bottlenecks (a temporary side-benefit of recession in the Canadian and world economy), Australian drought, and American political difficulties with the USSR and China. Improving the capacity

TABLE 3-4
WORLD WHEAT EXPORTS[1] AND MARKET SHARES

Exporters	*1960-64 (average)*	*1965-69 (average)*	*1970-74 (average)*	*1975-79 (average)*	*1980-81*	*1981-82*	*1982-83[2]*	*1983-84[3]*
				Million tonnes				
U.S.	19.5	19.0	25.6	31.8	41.9	49.1	41.5	38.0
Canada	11.0	11.3	12.7	13.9	17.0	17.8	21.0	21.5
Australia	6.1	6.5	7.5	9.8	10.6	11.0	7.5	11.0
EEC	3.3	4.2	4.9	7.6	14.7	15.5	15.5	15.5
Argentina	2.6	3.4	1.9	3.9	3.9	4.3	8.5	6.5
Other	5.7	8.2	7.5	5.2	6.2	4.5	6.2	4.9
Total[4]	48.2	52.6	60.2	72.2	94.2	102.2	100.2	97.4
				Percent share of exports				
U.S.	40.5	36.1	42.5	44.0	44.5	48.0	41.4	39.0
Canada	22.8	21.5	21.1	19.3	18.0	17.4	20.9	22.1
Australia	12.7	12.3	12.5	13.6	11.2	10.8	7.5	11.3
EEC	6.8	8.0	8.1	10.5	15.6	15.2	15.5	15.9
Argentina	5.4	6.5	3.2	5.4	4.1	4.2	8.5	6.7
Other	11.8	15.6	12.5	7.2	6.6	4.4	6.2	5.0
Total	100.0	100.0	100.0	100.0	100.0	100.0	100.0	100.0

Notes: [1] July/June, excluding intra-EEC.
[2] Preliminary.
[3] Forecast.
[4] Total might not add because of rounding numbers.

Source: USDA, *Wheat: Outlook & Situation* (May 1983).

FIGURE 3-2
WORLD WHEAT EXPORT SHARES, 1982-83

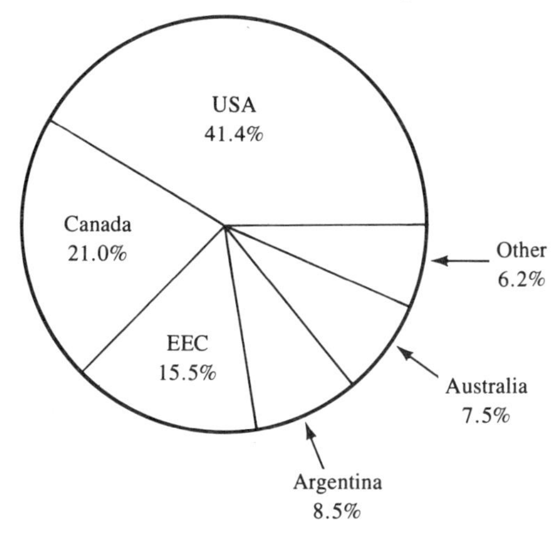

Source: Table 3-4

of Canada's grain handling and transportation systems is clearly a key factor in maintaining and improving Canada's relative position in world wheat exports.

Canadian production and export of wheat is dominated by hard red spring varieties, which command a market premium because of their relatively high protein content and other characteristics traditionally required by millers and bakers. Hard red spring varieties account for nearly 90 percent of the wheat produced in Canada and durum wheat, used for pasta products, accounts for much of the remainder. While the hard wheats are agronomically well suited to the semi-arid conditions that characterize much of the Canadian grain growing areas, changes in milling technology have reduced millers' dependence on these varieties, allowing more lower protein wheat to be used in bread flours. Consequently, higher yielding soft or medium hard wheats have captured a larger share of world trade in wheat, and the price premiums for hard red spring over other wheats are tending to narrow somewhat. Canadian export volumes of grain and grain products for five recent crop years are given in Table 3-5. The major countries where Canadian wheat was sold in 1981-82 are portrayed in Figure 3-3, which is based on Canadian Grain Commission figures that typically exclude wheat flour.

TABLE 3-5
CANADIAN EXPORTS OF GRAINS BY MAIN COUNTRIES OF FINAL DESTINATION
(thousand tonnes)

	1976-77	*1977-78*	*1978-79*	*1979-80*	*1980-81*	*1981-82*
Wheat (including durum)						
U.K.	1,431	1,525	1,312	1,360	1,409	1,366
Other Western	1,337	1,568	1,034	1,085	938	811
Poland	868	686	584	1,487	1,090	1,674
USSR	1,043	2,146	1,429	2,579	3,972	5,019
China	2,074	3,469	3,102	2,516	2,879	3,101
Japan	1,246	1,419	1,226	1,335	1,381	1,367
Brazil	1,031	782	948	1,270	1,284	1,314
Total Exports	12,709	15,240	12,299	15,212	15,567	17,972
Wheat Flour						
Cuba	479	454	483	423	326	370
Total Exports	734	757	762	614	692	474
Barley						
Western Europe	1,476	1,044	601	672	358	888
Eastern Europe	432	610	1,105	383	147	103
USSR	122	166	126	872	1,573	2,780
Japan	862	841	873	789	805	914
Total Exports	3,609	3,349	3,554	3,831	3,236	5,722
Rye						
Total Exports	168	271	154	397	446	547
Oats						
Total Exports	491	89	13	101	45	47

Sources: Canada Grains Council, *Canadian Grains Industry Statistical Handbook 1982*; and Canadian Wheat Board, *Annual Report 1981/82*.

Imports of wheat and flour are more widely dispersed among a number of countries than are exports. Over the 1960s and 1970s, the USSR was an erratic importer, sometimes of large amounts,[6] and the variability in these imports has been a substantial contributor to variability in world prices.[7] However, since the mid-1970s and particularly from 1979-80 to 1982-83, Soviet wheat imports have been rather less variable than formerly; this may be a reflection of the grain sale agreements that have been entered into by the USSR and major

FIGURE 3-3
CANADIAN WHEAT EXPORTS BY AREAS AND COUNTRIES, 1981-82

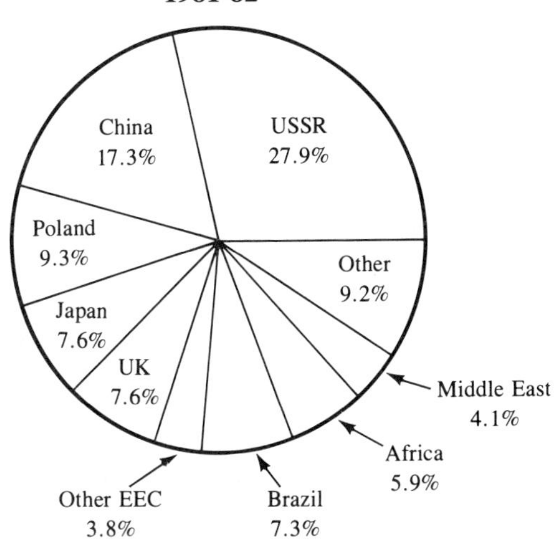

Source: Canadian Grain Commission data as reported in the Canadian Wheat Board, *Annual Report 1981/82*, pp. 9-10.

exporters during this period.[8] The Soviet share of world imports of wheat and flour was 16 percent in 1980-81.[9]

Other major importers of wheat are China, which accounted for nearly 15 percent of world imports in 1980-81, and Japan, with some 6 percent of imports in 1980-81. The import share of the EEC, which consistently declined over the 1960s and 1970s, was 5 percent in 1980-81.[10] Imports by developing countries and by centrally planned countries have grown rapidly, while the share of imports by developed countries has been declining. In 1971 developed market countries accounted for 81 percent of world imports of wheat and flour, but by 1981 they accounted for only 19 percent of world imports. In 1971 developing market countries accounted for 2 percent of imports, but by 1981 these countries accounted for 43 percent of imports. Centrally planned countries accounted for 16 percent of world imports of wheat and flour in 1971 and for 38 percent in 1981.[11]

World trade in flour is small compared to trade in wheat and currently accounts for only 10 percent of world trade in these products. Trade in flour has increased over time but not as rapidly as trade in wheat. Many importing nations have encouraged local processing of imported wheat and increased their own milling capacities. The USSR and Egypt are major importers;[12] Canada, Australia, the United States and the EEC have been the traditional major exporters. However, the EEC has substantially increased its share of the world export market for flour with the aid of export subsidies. EEC flour exports accounted for some 15 percent of world flour exports in 1953-54 and rose to nearly 70 percent by 1981-82.[13] This increased share of world flour exports was achieved mainly at the expense of two traditional exporters — Canada and Australia. The Canadian export market share was 30 percent in 1953-54 and fell to 6 percent by 1981-82, when the Australian market share had dropped to less than 2 percent. The market share of world flour exports from the United States, with the assistance of favourable credit terms offered to buyers, was much more stable at about 18 to 20 percent until the early 1980s.[14]

World Trade in Coarse Grains

World trade in coarse grains has grown more rapidly than trade in wheat, particularly during the decade prior to 1982 (see Table 3-2). Much of this increased growth in trade has been in corn. World trade in corn increased by 156 percent between 1971 and 1981, accounted for 68 percent of all coarse grains trade in 1981, and is dominated by the United States, which in 1981 accounted for 69 percent of world exports of corn. World trade in barley increased by 75 percent between 1971 and 1981 and accounted for 17 percent of world coarse grain trade in 1981.[15] In 1981 Canada accounted for 25 percent of world barley exports, France for 23 percent, the United States for 11 percent, the United Kingdom for 11 percent, and Australia for 8 percent.[16]

The majority of coarse grain imports are destined for developed countries; these nations accounted for 74 percent of world imports of corn and 75 percent of world barley imports in 1981. Japan and Western European countries are major importers: Japan accounted for 17 percent of corn imports and 8 percent of barley imports in 1981; Western European countries imported 27 percent of world corn imports and 32 percent of world barley imports. The USSR and Eastern European nations have imported increasing, though variable, quantities of coarse grains since the early 1970s; they accounted for 27 percent of corn and 32 percent of barley imports in 1981. China imports relatively minor quantities of coarse grains.

To sum up, less developed countries have become major importers of increasing quantities of wheat but account for a much smaller, albeit increasing, proportion of world coarse grain imports. Centrally planned nations are major importers in both these markets, although China primarily imports food grain wheat rather than coarse grains. Developed countries are major importers of increasing amounts of coarse grains but are becoming relatively less significant importers of wheat and flour. These changes in the structure of trade over the past two decades reflect two major factors: the influence of general growth in the world economy; and the nature and extent of national government policies towards domestic agriculture.

Economic and Politicial Factors Influencing Grain Trade

Gradual increases in GNP occurred in most of the Third World over the 1960s and 1970s. These increases, together with population increases, have raised the demand for food grains in less developed countries at a rate which, despite increased domestic agricultural production, has contributed to greater demand for food grain imports. Some substitution of wheat for traditional local food grains has accompanied urbanization and contributed to import demand for wheat in many Third World nations. As affluence increases, most people place more emphasis on improving the quality of their food intake by increasing protein intake. At high levels of income, consumption of food grains such as wheat is relatively unresponsive to price or income changes. Protein from animal sources — in the form of red meats and poultry and dairy products — tends to be emphasized, contributing to the demand for grains used in animal feed (feed grains) such as corn and barley and for other ingredients in livestock rations such as protein meal from vegetable oilseeds.

Another major influence on the trends and structure of world trade in grains and oilseeds arises from the widespread tendency of most governments — for political, social and economic reasons — to shelter part or all of their agricultural sector from the vagaries of world markets.[17] Restrictions on importation of agricultural products which compete with domestic products through tariffs, quotas or a variety of non-tariff barriers are common for grains and oilseeds, and for most agricultural commodities. Such practices often tend to insulate domestic prices from world market prices. Insulation of domestic prices from world market prices may also result from state-trading agencies, which are becoming increasingly important in import markets for grains and oilseeds. This is the case for the centrally planned economies, for many less developed countries and for some developed countries. The Canadian Wheat Board has calculated that, from 1977-78 to 1979-80, state-trading agencies accounted for about

81 percent of world imports of wheat and flour and for 58 percent of the world's coarse grain imports.[18]

Many high-income countries support part or all of their primary agricultural sector by price supports or subsidies to local producers. Again, grain and oilseeds are no exception. In some countries these programs are associated with extensive export subsidization of the resulting surplus of farm products. Insulation of domestic price levels for agricultural products such as wheat and coarse grains through various national programs of agricultural and trade intervention appears to force much of the burden of reaction to global changes in supply and demand on international markets. This accentuates the variability in traded volume and price levels in world markets. These tendencies are exemplified by the agricultural policy of the EEC. There is also a tendency towards greater politicization of world grain trade and this may also add to the instability and uncertainty in world grain markets. The response of the United States to the Soviet invasion of Afghanistan involving actions, initiated in January 1980 and which prevailed until the spring of 1981, to limit exports to the USSR to grain agreement levels provides only one example of this tendency.

One response to the uncertainty associated with increased instability in international markets for wheat and coarse grains since the early 1970s has been an increased emphasis on bilateral trade agreements by major grain exporting and importing nations. Some examples are the long-term grain trade agreements that both Canada and the United States have individually entered into with the USSR and China. These bilateral trading agreements generally relate to quantities of different types of grain to be traded; they provide umbrellas under which specific short-term sales contracts are negotiated. Nearly half of Canadian wheat exports are currently under long-term sales agreements.

Major grain trading nations may have tended to place increased emphasis on bilateral trade agreements partly because of the ineffectiveness of multilateral approaches. International negotiations, through the General Agreement on Tariffs and Trade (GATT), to reduce tariff and non-tariff barriers have had little impact on agricultural trade. An international commodity agreement intended to stabilize quantities and prices of grain traded internationally has been periodically renewed since 1948. However, the pricing provisions of this agreement — the International Wheat Agreement — broke down in 1968 under the pressure of world grain surpluses. The food-aid provisions of this agreement have been continued but the agreement no longer contains commercial trade provisions. Bilateral trading agreements are a rational mechanism for individual trading countries to reduce uncertainty of market access and supplies. However, several commentators have noted that if bilateral trade agreements continue to increase, they might have a destabilizing influence on a residual portion of world grain markets.[19]

The importance to world trade in grain of some major exporting and importing countries, together with the relatively small number of large private multinational trading firms (in particular, Cargill, Inc., the Continental Grain Company, the Louis Drefus Company, the Bunge Corporation, and André), export marketing boards (the Canadian and Australian wheat boards), and state-trading agencies, underly economists' application of imperfect competition models to explain price formation for grains in world markets. Following the Second World War and until the 1960s, the United States and Canada played a dominant role in controlling wheat prices, with Canada acting as a price leader.[20] This feature was recognized by McCalla,[21] who postulated a Canada-United States duopoly model, and by Alouze, Watson and Sturgess,[22] who proposed a triopoly model that also included Australia. These models imply the exertion of market power by major wheat sellers, a feature that has been disputed by Carter and Schmitz,[23] who postulate that the market power of wheat importers exceeds the market power of exporters. They suggest that such importers, particularly the EEC and Japan, impose an optimal import tariff on wheat. This view also underlies the argument, by some farm interest groups in the United States, that potential market power of that country could be bolstered by either a national marketing board or a cartel of major grain exporters.[24] This viewpoint also provides the background for a more extensive study[25] of the likely economic effects on major exporters of exerting countervailing power through an export cartel.

The alternative viewpoints of whether the balance of market power has been held by major exporters or by major importers of grain may each have been correct at different times. World markets for grains are not static but dynamic; their structure, including the nature and extent of government involvement, has changed over time; they are characterized by instability in volumes traded and in price levels. In some periods, as in 1973-74, there have been sellers' markets; in others, as in the late 1960s and, it appears, in the early 1980s, the balance of market power seems to have shifted to major importers.

International Markets for Canadian Oilseeds

Soybeans are the world's major oilseed crop, constituting more than half of world oilseed production.[26] Trade in soybeans and the products of soy oil and soy meal is dominated by the United States which, from 1975 to 1978, accounted for 81 percent of world exports of soybeans, 40 percent of soy meal exports and 32 percent of soy oil exports.[27] Brazil has also been a major exporter of these products since the early 1970s.

Rapeseed accounts for about 7 percent of world oilseed production. Canada accounts for nearly one-quarter of this share, and since 1978 it has been the major producer of rapeseed, closely followed by China and India. Canada is the major exporter of this oilseed, and in 1979 it accounted for 88 percent of world seed exports, 32 percent of oil exports and 52 percent of meal exports.[28] Domestic and international prospects for continued expansion of markets for Canadian rapeseed have been enhanced by the dramatic improvements in quality of rapeseed varieties achieved by Canadian plant breeders since the late 1960s. Oil from early rapeseed varieties contained high levels of erucic acid, a fatty acid with detrimental effects on the health of laboratory animals. The meal from early varieties contained undesirable levels of glucosinolates, chemical compounds that limited the use of rapeseed meal in livestock feed rations. In the 1970s rapeseed researchers produced varieties which combined low levels of both these substances. These "double-low" varieties are now called "canola" and these presently account for most Canadian rapeseed production.

Canadian flaxseed production averaged 22 percent of world production from 1975-76 to 1979-80. In 1979 Canadian exports of flaxseed were 97 percent of world exports but only a very minor proportion of world flaxseed oil and meal exports were from Canada. Export prices for Canadian oilseeds are largely influenced by prices for competing oilseeds: in 1981 rapeseed produced in Canada constituted less than 2 percent of world oilseed output. The world's major vegetable oilseed is soybeans, although cottonseed, peanuts, sunflower seed, rapeseed, palm, copra and flaxseed are also important vegetable oilseeds. Animal fats and oils from fish and olives are significant competitors for some oil and fat uses. Fish meal is a significant competitor in the meal market. Compared to soybeans, rapeseed has a relatively high oil content, and therefore prices for rapeseed tend to be more affected by prices in the vegetable oil market.

Nearly two-thirds of rapeseed traded in international markets is raw

TABLE 3-6
CANADIAN EXPORTS OF OILSEEDS
(thousand tonnes)

	1976-77	*1977-78*	*1978-79*	*1979-80*	*1980-81*
Rapeseed/Canola	1,018	1,014	1,720	1,743	1,372
Japan	757	777	1,017	1,073	1,147
EEC	216	33	425	462	110
Algeria	16	58	68	52	8
Rapeseed/Canola meal	107	156	170	176	204
Rapeseed/Canola oil	92	74	111	151	198
Flaxseed	356	263	494	455	519
EEC-9	195	143	346	273	323
Japan	78	96	90	111	111
U.S.	55	12	33	20	42
Soybeans	24	64	91	54	142
Sunflower seed	N/A	N/A	N/A	124	78

N/A = not available

Sources: Canada Grains Council, *Canadian Grains Industry Statistical Handbook 1982*, Table 27; and Statistics Canada, *Grain Trade of Canada 1980-81*, Table 70.

seed. The balance is traded as oil and meal. Japan is the dominant importer. These imports are primarily of unprocessed seed and amounted to nearly half of world seed imports in 1979. In that year, EEC imports amounted to 40 percent of seed imports, 84 percent of meal imports and 36 percent of oil imports. Algeria and India are consistent importers of rapeseed oil and also import some raw seed.

Canadian exports of oilseeds to major destinations are shown in Table 3-6. Canada has increasingly relied on Japan as the principal export market for rapeseed; in 1981-82 about 90 percent of Canadian exports of raw rapeseed went to that market.[29] India is the largest importer of Canadian rapeseed oil. West Germany, Norway, the United Kingdom and the Netherlands have generally been the major importers of Canadian rapeseed meal, while the United States has been an occasional meal importer.[30] The EEC countries are the major importers of flaxseed and flaxseed products. Japan is also a significant importer of raw flaxseed.[31] Canada has also been a periodic residual supplier of raw flaxseed to the United States.

Export and Production Prospects

4

The nature and strength of future export demand for grains and oilseeds are critical to the assessment of potential impacts on the Western and overall Canadian economy. The assessment of future export volumes, prices and values is strongly tied to a perception of the world food situation. Consequently, the question of export prospects is a matter of subjective judgment and considerable controversy. Like so many issues where crystal-ball gazing is involved, there is a high degree of uncertainty. This is particularly so with grain export prospects.

Export Prospects for the Grains Sector

In the early 1980s, it has been seen that relatively large and indeed record volumes of grain exports could coincide with lagging output prices, somewhat depressed farm incomes, and cost-price squeeze pressures, especially for new entrants to farming or highly leveraged farmers. Export prospects, however, should not be extrapolated from short-run circumstances — whether those might be the optimistic conditions existing between 1973 and 1975 or the more pessimistic market conditions prevailing in 1982 and 1983. It is medium- to longer-run world demand and supply conditions that must underlie realistic projections of world grain trade flows and probable grain export volumes and prices for Canada.

There have been several attempts to model future world production, consumption and trade flows in cereals.[1] Among the more recent studies are a Grains-Oilseeds-Livestock (GOL) model prepared by the U.S. Department of Agriculture[2] and the projection work of the International Food Policy Research Institute.[3] These models generally confirm the continuance of the historical trends shown in Table 2-3 and, in particular, the continued growth of grain trade deficits in less developed countries (LDCs). Indeed, the International Wheat Council has recently forecast that the developing nations will take over one-half of world grain imports by the year 2000.[4]

The basis for demand projections typically involves a constant price model in which population growth and income growth (the latter modified by estimated or assumed income elasticities of demand) are the two major demand shifters. For example, in the LDCs (including China), the current annual rate of population growth (denoted by p) is 2.1 percent,[5] the historical rate of increase per year in per capita income (y) is approximately 2 percent, and the income elasticity of demand for food (e) might be assumed to be 0.45. Given these values, the annual growth in the effective demand for food (f) in the LDCs can be projected to be 3 percent:

$$\begin{aligned} f &= p + ey \\ &= 2.1 + (0.45)\,(2) \\ &= 3 \end{aligned}$$

In the LDCs the historic rate of food production or supply increase has been approximately 2.5 percent per year over the past three decades. The essence of the world food problem, then, is not the popularly stated Malthusian concern that population growth is outstripping growth in food production in poor nations because, for the LDCs as a whole (although not for Africa), indices of per capita food production are increasing, albeit very slowly. Rather, the world food problem centres[6] on very serious problems of maldistribution of food (particularly associated with poverty and the lack of purchasing power by poor people in LDCs), together with the fact that cereal deficits in poor nations are steadily mounting since growth in the effective demand for food — generated by the combined impact of population growth and income growth — is outpacing growth in food supply.

The supply projections in global modelling generally assume a constant or very slowly growing land base, along with a continuance of historic yield increases into the future, and often contain no supply response mechanism by which quantity supplied is related to price. Indeed, the most general criticism that can be made of many projection models is that they appear to seriously under-estimate the productive capacity of rich nation agriculture. Despite the serious policy challenges raised by supply-related concerns in agriculture in rich nations — those associated, for example, with energy, pollution, water and land degradation — there has been a tendency by many observers in the past decade to over-rate these constraints on supply potential.[7] Furthermore, the longer-run projection models emphasize export/import volumes (placing more attention on this for food grains, especially wheat, and placing less emphasis on coarse grain prospects) and tend to be less clear on the question of prospective wheat and feed grain prices.

One of the more comprehensive and sophisticated of the global modelling efforts is the Grains-Oilseeds-Livestock (GOL) model developed by the U.S. Department of Agriculture in the mid-1970s, which was updated with projections to the year 2000 in the work for the Global 2000 Report to the U.S. President.[8] Unfortunately, the Global 2000 agricultural projections tend to over-estimate demand (especially for coarse grains, given the current slower shift to a diet oriented to more livestock products in most developed nations and several richer less developed countries), to under-estimate supply potential, to exaggerate trade flows and, in particular, to be misleading in projecting that the real price of food will increase 1 to 2 percent per year to the year 2000.

It is cause for some uneasiness that Agriculture Canada has partially based its strategy for agricultural and food sector development[9] on the Global 2000 forecasts without a more critical examination of those projections. Environment Canada also commissioned the study director of the Global 2000 Report to prepare a research report[10] outlining the implications of the Global 2000 Report for Canada. The Global 2000 projections suggest that Canadian wheat exports, while increasing 30 percent over the 1970-85 period, will decrease by 52 percent over the 1985-2000 period. This latter — and misleading — conclusion appears to stem from projection scenarios in which estimates of supply are overly constrained and those of domestic demand, assumed to be chiefly fuelled by increasing use of feed grains for red meat production, are over-estimated.

Most recent discussion on export prospects for the grains sector, however, has centred around the Canadian Wheat Board (CWB) export targets, which were announced in the later 1970s and updated in 1980 (see Table 4-1). The CWB export targets for Western Canada were 30 million tonnes of grain and oilseed exports by 1985 and 36 million tonnes by 1990. These export targets represented considerable increases over the range of grain exports (roughly 20 to 22 million tonnes) that prevailed at the time the targets were announced. It was also suggested in 1980 that 54 million tonnes of prairie grain production would be needed to meet the 1990 export target. In late 1982, the Canada Grains Council, in a more detailed study of prairie production possibilities and systems flows, concluded that a prairie

TABLE 4-1
THE CANADIAN WHEAT BOARD EXPORT TARGETS, 1990
(million tonnes)

Exports, Canada	*Range, 1990*	*Target, 1990*	*1978-79*
Wheat	20-24	22.0	13.0
Coarse Grains	7-10	8.5	4.1
Oilseeds	4-7	6.5	2.8
Total		36.0	19.9

Source: W.E. Jarvis, "Market Demand and Production Requirements for Prairie Grain," in Canada Wheat Board, Advisory Committee, *Prairie Wheat Symposium,* October 1980.

production target of 50 million tonnes was technically achievable and would be consistent with Canadian grain exports (implicitly of prairie origin) of 34 million tonnes.[11] (A 36 million tonnes export target would require the production of 52 million tonnes on the prairies to satisfy both export and domestic needs. The reduction of production requirements from 54 to 52 million tonnes appears to be associated with reduced need for prairie grain for domestic livestock feeding partly because of the increasing self-sufficiency of Eastern Canada in corn production.)

Record export movements during the past two crop years — 26.1 million tonnes in 1981-82 and the anticipated figure of some 28 to 29 million tonnes in 1982-83 — are remarkably in step with the CWB volume targets. These record export movements have occurred under conditions in which Canada has been fortunate to increase its share of a relatively stagnant world trade in grains since 1980. For example, grain transportation bottlenecks have not been serious due to recession in the Canadian and world economy and the associated reduced competition of grain with other resource products for Canada's limited transportation capacity, especially to the west coast. There was a very common tendency among many observers, including the CWB itself, to associate rising export volumes with more buoyant grain prices and incomes. In fact, record Canadian export volumes have been achieved under circumstances in the early 1980s in which world consumption has temporarily levelled off (in large part due to world recession), production and stock levels have been increasing, and price levels have been depressed.

Scrutiny of future demand/supply scenarios leads us to be sceptical of the view that the world is entering an era of chronic food scarcity and rising real food and grain prices. An educated guess would be that the international trade in grains will resume growth in the later 1980s (probably at 3 to 4 percent per year, or roughly half the unprecedented growth rates of the 1960s and 1970s), that Canada will have the opportunity to share in that expanding trade (chiefly to fill the growing cereal market in middle-income developing nations), but that real international grain prices are as apt to continue to follow a gradually falling trend (as they did for six decades prior to 1972) as to exhibit an increasing trend. Regardless of trend, there is fairly common agreement that grain prices will continue to be unstable.

Under these circumstances, does a Canadian grain export target of 34 to 36 million tonnes in 1990 appear realistic? In Figure 4-1, realized grain export volumes in the 1977-78 through 1981-82 crop years are plotted. Using the volume figure for 1981-82 of 26.1 million tonnes as a base, potential Canadian grain export levels throughout the rest of the 1980s are projected assuming an annual compound growth rate of 3 percent. (Such an annual growth rate is in line with the conclusion that world grain exports are likely to increase by some 3 to 4 percent per year.) Under these assumptions, the projected level of grain exports for Canada in 1990-91 is 34.1 million tonnes — a level that is virtually identical to the slightly revised target of the Canada Grains Council. The base year level assumed (26.1 million tonnes) is considerably above the levels of the previous four years. On the other hand, the preliminary export volume for 1982-83, estimated at more than 29 million tonnes by the Canadian Wheat Board, is well above the projected trend line. The International Wheat Council has recently forecast that world grain trade will increase by some 27 percent between 1980 and the year 2000.[12] Such an increase implies an annual compound growth rate of only 1.2 percent, which is certainly indicative of more pessimistic views of potential expansion in grain trade (see the lower projection trend line in Figure 4-1).

There are some concerns about the medium-run strength of export demand in both the USSR and China, Canada's two leading grain customers at the moment. The surge in Canadian grain sales to the USSR in the early 1980s (see Figure 2-2) was partially the result of four consecutive bad harvests in the Soviet Union. Future grain sales to the USSR are heavily conditional on the degree to which the USSR improves its agricultural production performance and the degree to which meat production and consumption is increased, both very problematic areas. A recent report by the Organization for Economic Co-operation and Development[13] concludes that the Soviet Union is

FIGURE 4-1
CANADIAN GRAIN EXPORTS: ACTUAL LEVELS, 1977-78 TO 1981-82, AND PROJECTIONS TO 1990-91 UNDER TWO ALTERNATIVE GROWTH SCENARIOS

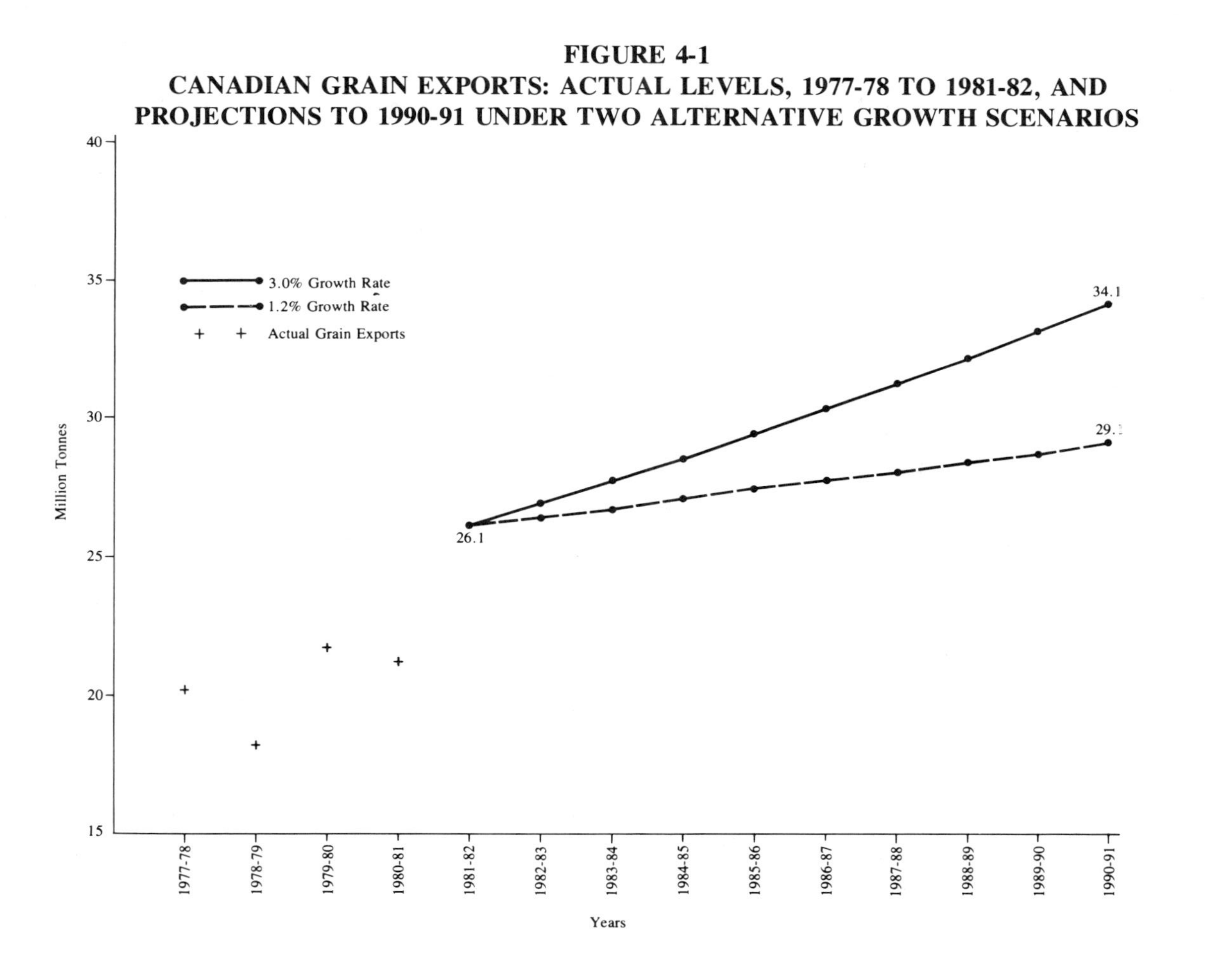

likely to reduce its dependence on agricultural imports over the balance of the present decade. This report suggests that USSR grain import requirements could fall to about 15 million tonnes by 1985 and fall below 10 million tonnes by 1990. (Canada alone currently provides the USSR with approximately 8 million tonnes of grain.) China is not expected to greatly increase or decrease current levels of grain imports through the 1980s.[14] Although China is believed to have a considerable but latent potential demand for feed grains, China's import demand for these grains is not likely to reach substantial proportions before the end of the 1980s. Overall, the export projections of the Canada Grains Council, which are generally consistent with the original Canadian Wheat Board targets, presently appear to be plausible, upper-range goals for policy and for future plans related to the grains sector. Higher levels of grain exports are not impossible, but would likely require some resurgence of world grain trade to the expansionary rates of the 1970s (an event not foreseen in late 1983) or an increase in Canada's market share (a phenomenon that is occurring in 1983, but mainly due to transitory factors), as well as continued improvement in Canada's grain transportation system.

Production Prospects to 1990

The achievement in 1990 of grain exports totalling 34 million tonnes would necessitate the production of 50 million tonnes of grain (including oilseeds) on the prairies. The Canada Grains Council has conducted a detailed research study of the technical and physical feasibility of attaining the 50 million tonnes goal by 1990.[15] Much as the Prairie Production Symposium concluded in 1980, such a goal is technically feasible, particularly if summer fallow acreage can be reduced and more extended cropping rotations can be introduced in the higher precipitation regions of the prairies.

The production increase that is required over base level production (the five-year average between 1977 and 1981) is some 13 million tonnes (see Table 4-2). Wheat production remains dominant and is projected to increase from 19.5 million tonnes to 27 million tonnes; this level was actually reached with the record wheat crop in 1982. Barley production is anticipated to rise from the 1977-81 average of 10.2 million tonnes to 15 million tonnes by 1990; the largest barley crop to date was 12.9 million tonnes, again in 1982, a bumper crop year. Canola (rapeseed) production is projected to increase from 2.6 to 3.8 million tonnes over the projection period.

The envisaged increases in grain and oilseed production are forecast

TABLE 4-2
CURRENT AND PROJECTED PRAIRIE PRODUCTION OF MAJOR GRAINS AND OILSEEDS
(million tonnes)

	1977-81 Average	*1990 Projection*	*Projected Increase*
Wheat	19.5	27.0	7.5
Oats	2.6	2.3	−0.3
Barley	10.2	15.0	4.8
Flaxseed	0.6	0.7	0.1
Canola	2.6	3.8	1.2
Total	35.5	48.8	13.3

Notes: Since the five major grains and oilseeds constitute 97 percent of total grains production, total grains and oilseeds production is projected to be 50.3 million tonnes in 1990.

This production target of 50 million tonnes (which includes the impact of a minor crop mix allocation shift towards oilseeds) is consistent with an export target of 34 million tonnes of grains, oilseeds and derived products from the prairies.

Source: Adapted from Canada Grains Council, *Prospects for the Prairie Grain Industry 1990* (1982), p. 114.

to arise from three major sources: 21.3 percent from new lands, 48.5 percent from reduced summer fallow acreage and 30.2 percent from increased yields. On a provincial basis, the projected increase of 13.3 million tonnes is expected to occur as follows: Manitoba, 2.1 million tonnes; Saskatchewan, 6.5 million tonnes (of which 4.8 million tonnes would be wheat); and Alberta, 4.7 million tonnes (of which 2.6 million tonnes would be barley). As a general rule, production increases are expected to be concentrated in the black soil zone or parkland regions of the prairies where moisture is more assured, chemical fertilization has a better payoff, and summer fallow can be more readily reduced in dryland production systems.

The 1990 production targets seem obtainable. Given excellent growing conditions in 1982, the estimated 1982 grain crop of 45.8 million tonnes (for the five major grain and oilseeds listed in Table 4-2) came relatively close to the 1990 goal. Economic circumstances, more so than physical constraints, are apt to play a larger role in whether the 1990 production targets are in fact achieved. We anticipate that prairie farmers will have some economic motivation to expand production, but this will occur as a result of output effects (the ability to move and sell larger volumes of grain) rather than price effects (if real world grain and oilseed prices show little or no increase or even decline slowly to the end of this century). Clearly, stronger price levels would elicit even higher levels of production, but we expect that the considerable productive capacity of rich nation agriculture (particularly American agriculture) will continue to dampen the possibilities of secular increases in real world grain prices.

The Economic Impact of the Grains and Oilseeds Sector

5

An expanding grains and oilseeds sector can be expected to have modest, but nevertheless important, impacts on the Western Canadian and overall Canadian economies. Clearly, the agricultural sector today will not provide the propulsive effects on Canadian economic development that accompanied the opening of the wheat economy in the period between 1900 and 1930. On the other hand, spinoff economic activities from resource- or staple-based industries such as the grains economy — despite the modest nature of linkage impacts associated with resource-based industries — remain important sources of growth to be exploited in the Canadian and prairie economies.

The Historic Role of Grains

The grains sector, specifically the wheat economy, has been regarded as "the keystone in the arch of Canada's National Policy." The National Policy, formulated in the years following Confederation in 1867, encompassed the set of programs to support the settlement of Western Canada, to build an all-Canadian railroad, and to protect Canadian industry.[1] Although the settlement of the Canadian West proceeded relatively slowly, the first three decades of the twentieth century were characterized by heavy immigration and rapid settlement, the establishment of the wheat economy, and an attendant investment boom in Canada. Between 1901 and 1921, for example, the population of the prairies, the number of prairie farms, and the output of wheat increased approximately five-fold.[2] The establishment and expansion of the wheat economy was a vitalizing force for the Canadian economy between 1900 and the onset of the Great Depression. Fixed capital investment was required for the settling of farms and the construction of support infrastructure for the burgeoning wheat economy. New settlers needed consumer goods as well as lumber and machinery to set up farms. The expanding wheat economy generated further economic activity, chiefly in central Canada but also significantly in the lumber industry in British Columbia, as economic spin-offs occurred in input supply industries, forward-linked transportation and handling sectors, and the protected consumer goods industries. Sir Wilfrid Laurier aptly described the grand design of economic policy in Canada at that time in a speech to the Canadian Manufacturers' Association in Quebec City in 1905:

> They [the settlers in Western Canada] will require clothes, they will require furniture, they will require implements, they will require shoes — and I hope you can furnish them to them in Quebec — they will require everything that man has to be supplied with. It is your ambition, it is my ambition also that this scientific tariff of ours will make it possible that every shoe that has to be worn in those prairies shall be a Canadian shoe; that every yard of cloth that can be marketed there shall be a yard of cloth produced in Canada; and so on and so on. . . .[3]

The wheat boom clearly increased the level of aggregate income in Canada and probably also led to higher levels of per capita income.[4] By 1930, however, the wheat economy had been established and extended, and its stimulative role was largely overtaken by newer staples — initially, newsprint and minerals and, later, oil and gas.[5] To what degree might an expanding grains sector in the later 1980s and 1990s reassert some of the growth-propulsive impacts with which it is historically and rightfully associated? There is no easy answer to this question. The grains and oilseeds sector in Western Canada exists today in a much different economic context than that which prevailed in the first three decades of this century. This may be attributed, in large measure, to the massive structural change in the Canadian economy that has accompanied the process of economic growth.

Structural Change and the Declining Role of Agriculture

As the economic development of rich industrial nations like Canada proceeds, structural change in the economy occurs in two major economic dimensions with respect to the agricultural sector. First, as per capita income levels have increased, the agricultural sector in Canada has contributed successively smaller proportions of overall output (either in terms of Gross National Product or Gross Domestic Product). In recent years, for instance, the agricultural sector's share of overall GDP in Canada has fallen to only 3.4 percent. Secondly, the declining relative role of the agricultural sector is evidenced in the sphere of employment. In 1982 agriculture's share of the employed labour force in Canada had dropped to only 4.4 percent.

This declining relative role of the agricultural sector is part and parcel of — and, indeed, is causally related to — the process of economic growth. On the demand side, the influence of Engel's Law (the empirical generalization that as income levels of households increase, the relative proportion of income spent on food decreases) is pervasive. Engel's Law is merely a reflection of differential income elasticities of demand for agricultural goods as opposed to industrial goods — in particular, that the domestic demand for agricultural products is not very responsive to upward shifts in income because the income elasticity of demand for agricultural goods is considerably less than 1 (it is often assumed to be about 0.2 or less in Canada and the United States). As a result, the output mix of a growing economy tends over time to be composed of a relatively smaller proportion of agricultural goods. Only the strength of foreign export demand can partially temper (or, perhaps, in short-time periods, reverse) this general tendency. While farm organizations and politicians sometimes bemoan the declining relative role of agriculture, this aspect of structural change is really a reflection of the growth of the Canadian economy and the enormous productive capacity of the agricultural sector. The fact that one Canadian farmer can produce food for twenty-two other Canadians — plus a sizeable exportable surplus — is a sign of the strength, not the weakness, of Canadian agriculture.

Even in Western Canada, the relative role of the agricultural sector has declined dramatically from its heyday during western settlement. As the wheat economy matured, the agricultural sector accounted for some two-thirds of the net value of production on the prairies from 1927 to 1929.[6] The agricultural share of economic activity was highest in Saskatchewan at 85.2 percent, followed by 69.8 percent in Alberta and 49.5 percent in Manitoba. By 1980, only 14.3 percent of GDP in the goods producing industries (including a few, but omitting most service sectors) is generated in the agricultural sector on the prairies.[7] Even in Saskatchewan, the agricultural sector now generates only one-third (34.0 percent) of goods production. In Manitoba, the agricultural share of goods production has dipped to 12.3 percent. In Alberta, the agricultural sector accounted for only 9.2 percent of goods production in 1980, although this percentage could increase temporarily in the 1980s as the oil and gas sector becomes somewhat less dominant.

It should also be noted that the crops sector in Canada has tended to decline relative to the livestock sector. During the period from 1926 to 1930, for instance, the crops sector comprised 53.7 percent of total cash receipts in Canadian agriculture;[8] during the past three decades the share of total cash receipts from crops (predominantly, but not solely, grains and oilseeds) has fluctuated between 38 and 46 percent, depending on the strength of foreign export demand.[9]

The process of structural change in the course of Canadian economic growth has led to the current situation in which the agricultural sector accounts for only small fractions of Canadian GDP and employment — 3.4 and 4.4 percent, respectively. More than any other factor, this circumscribes the degree to which potential expansion in the grains and oilseeds subsector can provide growth and investment opportunities in the overall economy.

Despite the declining relative role of the agricultural sector, there are important ways in which grains expansion can provide exploitable growth opportunities. The grains sector not only provides direct contributions to output, employment and trade but also generates indirect contributions through linked industries. While multiplier impacts associated with resource-based industries are typically not that strong, the nature of the Canadian economy is such that much of current and prospective economic growth is associated with primary resource sectors and primary or resource-based manufacturing. Furthermore, successful future economic diversification of the prairie economy is likely to be based on sectors — such as oil and gas (especially non-conventional sources) and agriculture — in which the West has a comparative advantage, rather than in artificial stimulation of secondary manufacturing activity.[10]

The Direct Contributions of the Grains Sector: Output, Employment and Trade

In the first place, of course, an expanding grains and oilseeds sector would make direct contributions to output growth, employment generation, and trade expansion. From a Canadian macroeconomic perspective, the trade impacts are the most important.

The Canada Grains Council targets of 50 million tonnes of production and 34 million tonnes of exports by 1990 imply that production levels would increase nearly 40 percent, and bulk export levels 60 percent, over base period (1977-81 average) levels.[11] The (real) gross value of grains and oilseeds production is projected to increase from 8.4 billion in 1981 to $10.2 billion (constant 1981 dollars) in 1990.[12] Grains expansion cannot be counted upon to have a major impact on Canadian value-added overall, given the relatively small size of the sector in the Canadian economy (the current share of Canadian GDP that is generated by primary grains and oilseeds production is only slightly more than 1 percent).

The impact of grains expansion on employment should be considered in terms of slowing the rate of exodus of workers from the agricultural sector. During the past decade, the absolute number of workers in agriculture has stabilized at about 500,000, although the relative share of the employed labour force in agriculture has continued to decline. There are many factors — for example, technological change, associated capital-for-labour substitution and part-time farming — that complicate the forecasting of future levels of employment in agriculture. Grain farming technologies in the 1980s are very capital-intensive and output expansion will not have major effects on direct labour employment in grain and oilseed production. However, on the whole, grains expansion can be anticipated to slow the historic outflow of labour from agriculture and to slow the rate of attrition in the number of farms.

Given record-level volumes of grain exports in the early 1980s, the grains and oilseeds sector in Canada is currently making very important contributions to the growing surpluses in merchandise trade in agricultural products. This, in turn, is one positive factor, among many other positive and negative factors, impinging on Canada's overall current account trade balance and the strength of the Canadian dollar. In 1981 the value of gross exports of cereal grain and grain products was $5.2 billion whereas the value of gross exports of oilseeds and oilseed products was $1.0 billion; together they comprise approximately 70 percent of the value of all Canadian agricultural exports (see Table 2-3). Canada's net trade balance in cereal grains and oilseeds was a $5.3 billion surplus ($4.8 billion from cereal grains and grain products and $0.5 billion from oilseeds and oilseed products). This large net trade surplus for grains and oilseeds was a major factor in Canada's achievement of an overall net trade surplus for all agricultural commodities of $3.2 billion (see Table 5-1). In 1982 the agricultural trade surplus is estimated to have reached a record $4.2 billion dollars ($9.3 billion of agricultural exports less $5.1 billion agricultural imports).[13] While this record-level agricultural surplus is partially due to a 10 percent drop in imports of agricultural commodities in 1982, it is also due to record export levels for wheat (exceeding $4 billion for the first time) and barley ($0.9 billion).

TABLE 5-1
CANADIAN NET TRADE BALANCES FOR AGRICULTURAL COMMODITY GROUPS[1]
(millions of current dollars)

	1971-75 Average	*1976-80 Average*	*1981*	*1982*[2]
Grains and grain products	1,726	2,848	4,827	5,287
(Wheat and wheat flour)	(1,411)	(2,326)	(3,723)	N/A
Oilseeds and oilseed products	121	266	498	365
Animal feeds	46	98	135	136
Animals, meat and other animal products	6	205	400	799
Dairy products	12	32	115	184
Poultry and eggs	−5	−33	−17	−33
Fruits, nuts and vegetables	−504	−1,030	−1,427	−1,480
Other	−551	−1,110	−1,362	−1,010
Total net trade in agricultural commodities	851	1,276	3,169	4,248
Net trade in all commodities	382	2,050	1,855	13,903

Notes: [1] For agricultural trade balances in 1951, 1961, 1966, 1971 and 1976 calculated by this classification scheme, see B.W. Wilkinson, *Canada in the Changing World Economy* p. 98.
[2] Preliminary.

Source: See Table 2-3.

Clearly, when Canadian agricultural exports comprise some 11.4 percent of all Canadian commodity exports, the role of agriculture in overall Canadian trade performance should not be exaggerated. The dominant factor in Canada's very large trade surplus of $13.9 billion in 1982 is the drastic recession-induced drop in imports. Nevertheless, the contribution of grains and oilseeds to the Canadian current account should not be overlooked. During the 1980s, it can be anticipated that bulk grain and oilseed exports will grow by approximately 1 million tonnes per annum. Given an average value of $200 per tonne, the (real) value of grain and oilseed exports might be expected to increase by at least $200 million per year. One would hope that this might be one factor, among many, that might assist in moving the Canadian economy toward recovery and improved economic growth. Increased grain and oilseed exports are a factor which contributes to improved merchandise trade performance, less strain on the Canadian balance of payments, less downward pressure on the Canadian dollar, and less need for restrictive monetary policy.

The Indirect Impact of the Grains Sector on Linked Industries

In addition to direct impacts, the grains and oilseeds sector also generates economic activity indirectly. The activities of processing, handling and transporting occur in industries forwardly linked to the grains and oilseeds sector. The main backward linkages are associated with industries that supply inputs for primary crop production. Final demand linkages are generated in industries that supply consumer goods to grain farmers. In this section, a general overview of these various linked industries is provided; a more detailed examination of the nature of problems and prospects associated with specific forward-linked industries is given in Chapter 6, while key backward-linked industries are discussed in Chapter 7.

The main forward- and backward-linked industries associated with primary grain and oilseed production in Canada are cited in Table 5-2. The most important area for further use of grains and oilseeds is livestock feeding, and economic activity generated in the feed milling sector is the most significant forward linkage. In 1981, the feed milling industry in Canada — some 600 establishments — produced shipments valued at $2.5 billion, employed nearly 10,000 workers, and generated value-added of $472 million. The most important areas of further processing include the milling of wheat flour, the malting of barley and the crushing of rapeseed. Economic

TABLE 5-2: INDUSTRIES LINKED AND RELATED TO THE GRAINS SECTOR, 1981

	Establishments (number)	*Value of Shipments ($000)*	*Total Employees (number)*	*Total Salaries and Wages ($000)*	*Total Value-Added ($000)*
Backward-linked Industries					
Agricultural implement	207	1,402,504	16,073	342,686	686,447
Primary fertilizer	30	N/A	10-12,000[1]	N/A	N/A
Mixed fertilizer[2]	105	218,443	1,017	18,419	67,936
Pesticides	N/A	320,000[3]	N/A	N/A	N/A
Forward-linked Industries					
Feed[4]	601	2,524,255	9,683	174,742	471,756
Flour & breakfast cereal products	50	1,111,852	5,214	113,449	346,153
Vegetable oil mills	11	829,029	1,525	35,784	129,236
Related Industries					
Slaughtering & meat processors[5]	501	7,603,018	35,450	706,004	1,294,787
Bakeries	1,431	1,358,022	26,347	438,222	744,795
Biscuit manufacturers	28	411,845	6,371	110,350	244,859
Breweries[6]	40	1,444,113	12,637	334,316	1,003,462
Distilleries[7]	33	782,096	5,528	132,112	484,077
Food & Beverage Industries[8]	4,492	31,841,656	234,077	4,362,989	10,354,470

Notes: [1] Estimated. [2] *Statistics Canada Daily*, 18 March 1983, advance information. [3] This figure is an estimate of the value of shipments of herbicides, crop and seed treatments, and livestock pesticides for the year ending 30 September 1981.

[4] *Statistics Canada Daily*, 29 April 1983, advance information. This industry consists of establishments primarily engaged in producing balanced feeds and premixes or feed concentrates. The feeds contain ground or rolled grains, mill feeds, animal and vegetable proteins, minerals, essential vitamins and antibiotics. Local mixing and custom grinding of grains are included in this industry.

[5] *Statistics Canada Daily*, 29 April 1983, advance information. [6] *Statistics Canada Daily*, 15 October 1982, advance information.

[7] *Statistics Canada Daily*, 8 October 1982, advance information. [8] Preliminary figures from *Statistics Canada Daily*, 24 June 1983, ''1981 Census of Manufacturers — Selected Principal Statistics of the Manufacturing Industries of Canada, by Major Group.''

Sources: Statistics Canada, *Agricultural Implement Industry*; V.A. Heighton, ''Agricultural Chemicals and Other Supplies'' in *Market Commentary, Farm Inputs and Finance* (Agriculture Canada, December 1981), Table 1; Statistics Canada, *Feed Industry*, Cat. 32-214; Statistics Canada, *Flour and Breakfast Cereal Products Industry*, Cat. 32-228; Statistics Canada, *Vegetable Oil Mills*, Cat. 32-223; Statistics Canada, *Bakeries*, Cat. 32-203; and Statistics Canada, *Biscuit Manufacturers*, Cat. 32-202.

activity in flour milling has tended to decline slowly over time, while rapeseed crushing and barley malting, both starting from small bases, evidenced strong growth in the last decade.

The agricultural sector has become increasingly reliant on off-farm inputs supplied by backward-linked sectors. Farmers in the prairie provinces in 1982, for example, spent $685 million on machinery repairs, $843 million on petroleum products, $607 million on fertilizer and $335 million on pesticides (see Table 5-3). Farmers' estimated depreciation expenses, machinery being a key item, totalled $1,662 million. The largest backward-linked industry is the agricultural implement industry with shipments valued at $1.4 billion, 16,000

TABLE 5-3
TOTAL FARM OPERATING AND DEPRECIATION EXPENSES,[1] PRAIRIE PROVINCES, 1982

	Prairie Provinces ($ millions)	*Share of Total Operating & Depreciation Expenses (%)*
Property taxes	147.2	1.9
Gross farm rent	512.5	6.6
Wages to farm labour	389.1	5.0
Interest on debt	1,055.3	13.6
Petroleum products	842.7	10.8
Machinery repairs	684.5	8.8
Fertilizer	606.9	7.8
Pesticides	335.1	4.3
Feed	451.4	5.8
Repair to buildings	116.8	1.5
Total Operating Expenses[2]	6,121.1	78.6
Depreciation	1,662.3	21.4
Total Operating & Depreciation Expenses	7,783.4	100.0

Notes: [1]There is no imputation of costs for owner-operator and non-paid family labour, and capital services of owned land.
[2]Includes all other farm operating expenses.

Source: Agriculture Canada, *Selected Agricultural Statistics: Canada and the Provinces, 1983*.

employees, and value-added of $686 million in Canada in 1981. There are significant leakages of input expenditures on imported goods in the case of agricultural machinery and pesticides (as will be shown in Chapter 7).

The third general area in which the grains economy induces further economic activity results from expenditures of factor incomes earned in grains and oilseeds production on consumer goods. Unlike other resource-based industries, there are very limited leakages of these factor incomes, at least in terms of capital servicing, outside the country. However, given the relatively small proportion of the Canadian population engaged in grains and oilseeds production (only 5.5 percent of the Canadian population resided on farms in 1981 and only about one-third of the farm population can be associated with grains and oilseeds production) and given the capital-intensive nature of grains production and the expected minimal increase in direct labour employment, grains expansion can be anticipated to have only minor impacts in inducing greater economic activity in consumer goods industries in Canada. However, in prairie cities and larger towns, the economic health of the grains sector does have more major impacts on retail sales and the prosperity of the local business community. As noted in Chapter 3, while grain production and exports may increase in the next decade, there are serious reservations about whether grain prices will rise and whether grain net incomes will necessarily be consistently buoyant.

Assessment of the Overall Impact

An indication of the economic impact of the grains and oilseeds industry can be derived from the fact that the primary producing agricultural sector (that is, the farm sector itself) has contributed some 3.4 percent to overall GDP in Canada in recent years.[14] Given that total cash receipts for grains and oilseeds are about one-third of total cash receipts in Canada[15] (and assuming that the grains and oilseeds sector's value-added share is roughly proportional to its cash receipt share), one can deduce that grains and oilseeds directly contribute about one-third times 3.4 percent which is equal to 1.13 percent of Canadian GDP. Furthermore, using multiplier analysis and assuming a gross output or impact multiplier of 1.76 for the grains sector,[16] the direct and indirect requirement of inputs to generate such a level of deliveries to final demand for grains can be derived by multiplying 1.76 by 1.13 which equals 1.99 or approximately 2 percent of Canadian GDP.

This rough estimate, while counting backward-linkage impacts associated with input supply industries, does not include forward-linkage effects such as those associated with processing or transportation and handling. Evidently, since potential grains expansion would predominantly be associated with increased export demand in the form of bulk grains, the linkage effects would be greatly constrained and confined primarily to the extra economic activity generated in the transportation and handling industries as largely unprocessed product was moved into export position. Nor does the calculation include induced income impacts (such as those calculated when the household consumption sector is made endogenous in input-output analysis) as farm families spend their incomes on consumption goods. On the other hand, the usual caveat or warning about multiplier analytics is in order; the addition of multiplier impacts across all industries would lead to the nonsensical result that double or triple the actual GDP level was being accounted for.

In a document published in 1983, *The Role and Importance of Grains and Oilseeds to the Canadian Economy*, the Canada Grains Council estimates that the grains and oilseeds sector (defined more widely than merely the producing sector to also include forward-marketing chain activity) contributes about 5 percent of Canadian GDP at factor cost.[17] The authors of this study estimate the value of grains, oilseeds and derived products as close to their final demand use as possible. The major final demand uses considered are: food ($4.0 billion); alcoholic beverages ($1.7 billion); imputed value of farm feed ($3.1 billion); net export demand ($5.6 billion); and non-agricultural industrial uses (not estimated due to lack of data, but regarded to be minor). These uses were calculated to total nearly $14.5 billion of final demand or some 5.4 percent of Canadian GDP in 1980. While there is a wealth of empirical information and several useful insights in this research study, the methodology employed overstates the relative impact of the grains and oilseeds sector. The primary shortcoming is the attribution of all value-added in marketing chain activity solely to the grains sector when grains, an intermediate input, is only one of several inputs employed in these forwardly linked production processes.

Oil and Wheat: Leading Sectors?

There is a tendency among economists, especially when looking at primary resource sectors, to be rather pessimistic about the multiplier impacts of these sectors. Indeed, Caves and Holton, in assessing the economic impact of the oil and gas boom in Alberta in the decade following the Leduc discovery in 1947, concluded that "the local repercussions of the petroleum and gas development are likely to be distinctly limited."[18] This was due to such factors as the capital-intensive production technology and rather limited direct employment impacts of the sector, the associated leakages with respect to both capital servicing (interest and dividend payments to foreign investors) and imported inputs, and the limited strength of the linkages of the petroleum sector. These factors, except the leakages associated with capital servicing, also tend to characterize the grains sector.

Owram contends that with hindsight Caves and Holton would change their conclusion.[19] Perhaps a compromise between these two perspectives would be to view the record of the petroleum sector in inducing growth and structural transformation of the prairie economy, especially the Alberta economy, as mixed.[20] On the one hand, the oil and natural gas sector experienced rapid growth in the value of its output in both money and real terms, although this growth between 1973 and 1981 was mainly a result of price rather than output effects. Moreover, the petroleum sector in Alberta assisted indirectly in inducing activity in a variety of related industries, including construction and services, and large economic rents from the sector provided the wherewithal for capital spending on social and physical infrastructure. On the other hand, during the past two decades in Alberta, there has been a surprising lack of change in the sectoral composition of non-agricultural employment (the manufacturing share, in fact, showing a slight decline) and in the industrial composition of manufacturing employment (with meat slaughtering and packing remaining the leading manufacturing subsector).

We are also inclined to view the impact potential of grains expansion as being mixed. Many of the constraining features, as outlined by Caves and Holton, are clearly present. Moreover, the grains and oilseeds sector is now a relatively small share of the Canadian economy, and indeed even of the prairie regional economy. Nor does the grains sector have the direct rent-generating capacity of the oil and gas sector. On the other hand, resource-based sectors such as petroleum (particularly heavy oil and tar sands development) and grains probably present the strongest foundation for future development and diversification of the prairie economy. Even for the overall Canadian economy, primary resource extraction and primary resource manufacturing remain relatively important spheres of economic activity. An examination of impact multipliers for various sectors of

TABLE 5-4
IMPACT MULTIPLIERS FOR SELECTED CANADIAN COMMODITIES, 1979

Commodities	*Multipliers*
Grains	1.76
Forestry products	1.97
Fish landings	1.40
Coal, oil and gas	1.66
Meat products	2.21
Dairy products	2.27
Feeds	2.25
Flour, meal and cereals	2.27
Textiles	1.81
Lumber and plywood	2.13
Iron and steel products	1.91
Agricultural machinery	1.68
Motor vehicles	1.60
Fertilizers	1.74
Transportation margins	2.66

Source: Statistics Canada, *The Input-Output Structure of the Canadian Economy 1971-79* (1983), Table 62 (1979 Impact Matrix; Aggregation M).

the Canadian economy reveals, somewhat surprisingly, that the impact multipliers for grain and its related sectors are about as favourable as one finds in the Canadian economy (see Table 5-4). This is not to argue that recovery and expansion of the Canadian economy in the rest of the 1980s can ride on the coat-tails of moderate grains expansion, but merely to point out that the grains sector presents worthwhile growth opportunities which should not be neglected.

The Processing and Use of Canadian Grains: The Forward Linkages

6

The supply and disposition of major Canadian grains and oilseeds from the mid-1970s to the early 1980s are indicated in Tables 6-1 and 6-2. These tables show that, overall, wheat is primarily sold in export markets while coarse grains are mainly used for domestic purposes. Exports do not greatly exceed domestic use of rapeseed/canola although they represent the major market outlet for Canadian flaxseed. Imports are an appreciable and relatively stable proportion of total soybean supplies. Grain and oilseed exports consist predominantly of bulk products; processed products constitute a relatively small proportion of grain and oilseed exports.

Table 6-3 provides information on the various categories of domestic use of major Canadian grains and oilseeds. More detailed information concerning the extent of Canadian processing of grain and oilseed products and the volumes exported is given in Table 6-4. The major use for coarse grains — and the major domestic use for all grains — is as animal feed. This category (which also includes data for waste and dockage) has been remarkably stable in aggregate terms over the 1970s and early 1980s. With the exception of corn, the volume of grain fed to livestock has declined somewhat since the very early 1970s when depressed grain markets and sizeable farm stocks of grain encouraged grain feeding of livestock, particularly in the prairies.[1] Domestic human consumption of cereal grains in Canada is primarily in the form of flour and breakfast cereals. Canadian per capita consumption of these grain products has been relatively static since the 1960s.[2] Domestic processing of barley to produce barley malt has increased over the 1970s and early 1980s. Nearly half of the malt produced in Canada is exported (Table 6-4) and the balance is used domestically in breweries and distilleries. Canadian oilseed crushing activity has risen steadily throughout the 1970s and early 1980s, as has

TABLE 6-1
WHEAT, COARSE GRAINS, AND OILSEEDS SUPPLY, CANADA
(thousand tonnes)

Crops	*Beginning Stocks*		*Production*		*Imports*		*Supply*	
	5-Year Average[1]	*1981/82*	*5-Year Average*	*1981/82*	*5-Year Average*	*1981/82*	*5-Year Average*	*1981-82*
All Wheat	11,808.8	8,570.2	20,213.8	24,802.5	—	—	32,022.6	33,372.7
Coarse Grains								
Barley	3,618.2	3,203.2	10,516.9	13,724.2	5.8	1.5	14,140.9	16,928.9
Oats	1,328.8	759.6	3,694.5	3,188.4	14.3	12.0	5,037.6	3,960.0
Rye	366.9	222.1	487.4	927.2	6.0	—	860.3	1,149.3
Corn	1,009.0	1,273.8	4,703.6	6,673.4	817.1	822.2	6,529.7	8,769.4
Total	6,322.9	5,458.7	19,402.4	24,513.2	843.2	835.7	26,568.5	30,807.6
Total Grains	18,131.7	14,028.9	39,616.2	49,315.7	843.2	835.7	58,591.1	64,180.3
Oilseeds								
Rapeseed	823.5	1,327.9	2,440.3	1,836.7	—	—	3,263.8	3,164.6
Flaxseed	408.4	344.0	551.7	468.0	0.1	—	960.2	812.0
Soybeans	28.4	53.0	538.6	607.0	364.4	423.7	939.2	1,083.7
Total Oilseeds	1,260.3	1,724.9	3,530.6	2,911.7	364.5	423.7	5,163.2	5,060.3

Note: [1]Five-year average for 1976-77 to 1980-81.

Source: Statistics Canada, *Cereals and Oilseeds Review* (1983); and Canada Grains Council, *Canadian Grains Industry Statistical Handbook, 1982*.

TABLE 6-2
DISPOSITION OF MAJOR CANADIAN GRAINS AND OILSEEDS
(thousand tonnes[1])

	Total Disposition		*Grain Exports*		*Processed Product Exports*		*Domestic Disappearance*		*Ending Stocks*	
Crops	*5-Year Average*[2]	*1981/82*	*5-Year Average*	*1981/82*	*5-Year Average*	*1981/82*	*5-Year Average*	*1981/82*	*5-Year Average*	*1981/82*
All Wheat	32,022.6	33,372.7	14,208.6	17,973.2	733.8	473.6	5,153.2	5,168.1	11,927.0	9,757.8
Coarse Grains										
Barley	14,140.9	16,928.9	3,515.8	5,722.4	273.6	289.1	6,645.4	6,756.1	3,706.1	4,161.3
Oats	5,037.6	3,960.0	148.9	50.1	1.1	0.5	3,653.1	3,054.0	1,234.5	855.4
Rye	860.3	1,149.3	287.0	561.1	—	—	223.9	258.7	349.4	329.5
Corn	6,529.7	8,769.4	418.9	1,131.1	—	—	4,984.4	6,453.0	1,126.4	1,185.3
Total	26,568.5	30,807.6	4,370.6	7,464.7	274.7	289.6	15,506.8	16,521.8	6,416.4	6,531.5
Total Grains	58,591.1	64,180.3	18,579.2	25,437.9	1,008.5	763.2	20,660.0	21,689.9	18,343.4	16,289.3
Oilseeds										
Rapeseed	3,263.8	3,164.6	1,373.8	1,359.3	305.8[3]	398.0[3]	704.9	714.9	879.3	692.4
Flaxseed	960.2	812.0	425.3	447.7	17.8[3]	24.6[3]	116.0[4]	77.3[4]	401.1	262.4

Notes: [1]Products are expressed in terms of grain or oilseed equivalent weights.
[2]Five-year average for 1976-77 to 1980-81.
[3]Estimated as seed equivalent of oil exports.
[4]Adjusted for product exports.

Sources: Statistics Canada, *Cereals and Oilseeds Review* (1983); Statistics Canada, ''Supply and Disposition of Major Grains''; and Canada Grains Council, *Canadian Grains Industry Statistical Handbook, 1982*.

TABLE 6-3
DOMESTIC USE OF MAJOR CANADIAN GRAINS AND OILSEEDS
(thousand tonnes[1])

Crops	*Domestic Disappearance*		*Human Food*		*Feed, Waste, Dockage*		*Industrial Use*		*Seed and Loss*	
	5-Year Average[2]	*1981-82*	*5-Year Average*	*1981-82*	*5-Year Average*	*1981-82*	*5-Year Average*	*1981-82*	*5-Year Average*	*1981-82*
All Wheat	5,153.2	5,168.1	1,890.5	1,935.2	2,213.3	2,053.9	20.2	19.3	1,029.2	1,159.7
Coarse Grains										
Barley	6,645.4	6,756.1	6.5	8.1	5,842.2	5,866.1	364.8[3]	397.8[3]	431.9	484.1
Oats	3,653.1	3,054.0	69.8	74.7	3,408.3	2,813.7	—	—	175.0	165.6
Rye	223.9	258.7	13.5	14.0	114.6	143.4	67.8	60.0	28.0	41.3
Corn	4,984.4	6,453.0	836.2	1,122.0	4,125.3	5,305.4	[a]	[a]	22.9	25.6
Total	15,506.8	16,521.8	926.0	1,218.8	13,490.4	14,128.6	432.6	457.8	657.8	716.6
Total Grains	20,660.0	21,389.9	2,816.5	3,154.0	15,703.7	16,182.5	452.8	477.1	1,687.0	1,876.3
Oilseeds										
Rapeseed	704.9	714.9	455.3[4]	547.4[4]	230.6	151.7	[a]	[a]	19.0	15.8
Flaxseed	116.0[5]	77.3[5]	[b]	[b]	[b]	[b]	[b]	[b]	[b]	25.9

Notes: [a]Included with human food.
[b]Confidential to meet secrecy requirements of the Statistics Act.
[1]Products are expressed in terms of grain or oilseed equivalent weights.
[2]Five-year average for 1976-77 to 1980-81.
[3]Includes malt used in brewing and distilling.
[4]Crushings, adjusted for oil exports.
[5]Adjusted for product exports.

Sources: Statistics Canada, *Cereals and Oilseeds Review* (1983); Canada Grains Council, *Canadian Grains Industry Statistical Handbook, 1982*.

TABLE 6-4
SELECTED STATISTICS FOR DOMESTIC PROCESSING OF GRAIN IN CANADA
(thousand tonnes[1])

Grain		*1976-77 to 1980-81 Average*	*1981-82*[2]
Wheat	Processed	2,590.9	2,369.2
Flour	Produced	1,912.6	1,775.5
Flour	Exported	537.7	348.1
Millfeeds	Produced	673.6	563.9
Millfeeds	Exported	256.2	148.1
Oats	Processed	70.7	77.2
Oatmeal & Rolled Oats	Produced	36.4	38.6
Rolled Oats	Exported	0.6	0.2
Barley	Processed	625.7	675.5
Malt	Produced	465.2	502.5
Malt	Exported	213.0	243.7
Rye	Processed	13.0	13.3
Flour & Meal	Produced	10.4	10.9
Rapeseed	Crushed	761.1	945.3
Oil	Produced	312.8	382.1
Oil	Exported	125.1	162.8
Meal	Produced	436.7	551.1
Meal	Exported	162.7	162.4
Soybeans	Crushed	804.8	961.9
Oil	Produced	136.8	164.3
Oil	Exported	6.2	17.3
Oil	Imported	23.5	3.6
Meal	Produced	630.8	757.5
Meal	Exported	50.9	48.7
Meal	Imported	400.7	406.3
Flaxseed			
Oil	Exported	6.2	8.6
Meal	Exported	7.0	5.3

Notes: [1]Products are in actual weights rather than grain equivalents.
[2]Preliminary.

Sources: Statistics Canada, "Supply and Disposition of Major Grains" (1983); for barley and malt, see Canadian International Grains Institute, *Grains and Oilseeds Handling, Marketing, Processing* (1982).

TABLE 6-5
MAJOR END USES OF GRAINS[1] PRODUCED ON THE PRAIRIES, 1976-77 TO 1980-81, AND CANADA GRAINS COUNCIL PROJECTIONS OF USE IN 1990
(thousand tonnes)

	1967-77 to 1980-81 Average	*1990 Projection*	*Percent Increase*
Exports (bulk grain)	19,284	30,950	60.5
Fed within the Prairies	7,462	7,371	−1.2
Processed on the Prairies	2,173	3,540	62.9
Shipped as feed to British Columbia	327	390	19.3
Shipped to Eastern Canada	3,303	3,548	7.4
as feed	1,319	1,330	0.8
for other domestic uses	1,984	2,218	11.8

Note: [1]Includes wheat, barley, oats, rapeseed and flaxseed.

Source: Canada Grains Council, *Prospects for the Prairie Grain Industry, 1990*, p. 189.

refining of domestic vegetable oilseeds. Domestic consumption of oil and meal and exports of these products has also increased. Table 6-5 provides a summary, prepared by the Canada Grains Council, of the end uses of five major prairie grains and oilseeds. This table also gives the council's projections of the use of prairie grain in 1990. The following sections provide more detailed outlines of the various forward-linked processing industries together with a brief overview of the grain handling and distribution system.

Feed Grain Use

Livestock feed is the largest domestic use category for barley, oats and the other coarse grains, and accounts for appreciable quantities of feed wheat. By-products from the flour milling, malting and brewing industries, as well as by-products from grain cleaning, are also incorporated into animal feeds. The feed milling industry is the largest grains and oilseeds manufacturing industry. In 1981, the Canadian industry involved 601 establishments and 9,683 employees (see Table 5-3).

The products of the feed milling industry are nearly all consumed domestically. A very small proportion of Canadian grain and grain product exports are in the form of processed or semi-processed animal feeds. (Grain-based processed and semi-processed animal feed accounted for just 3 percent of the value of total grain and grain product exports in 1981.) A considerable quantity of feed grains used in Canada is not processed by commercial feedmills and does not enter commercial channels but is fed to livestock on the farm where it was produced or is sold to other farmers. Most of the oats and a large proportion of barley used in Canada by-pass commercial feed grain channels.

The regional distribution of feed grain consumption and feed milling activity within Canada tends to be similar to the distribution of livestock production across the country. The hog sector accounted for 35 percent of domestic feed grain use in 1981-82 followed by beef (29 percent), dairy (20 percent) and poultry (16 percent). The prairie region, considered as a whole, is the major feed grain consuming area, accounting for an estimated 38 percent of domestic use in 1981-82, followed by Ontario (27 percent) and Quebec (26 percent).[3] Traditionally, Eastern Canada and British Columbia have been grain-deficit regions, relying on inflows of grain from Western Canada and, to a lesser extent, on imported corn from the United States for livestock feeding and other uses. However, increasing production of feed grains outside the prairies has reduced the dependence of these regions on inflows of western grain. The prairies are the major grain producing region, accounting for 96 percent of wheat production and 65 percent of coarse grains production in Canada in 1981-82. Barley, the most important coarse grain, is grown mainly in the prairies. However, Canadian production of grain corn has increased steadily. Ontario is now a significant supplier of corn to Quebec and the Atlantic provinces. By 1981-82 corn accounted for 27 percent of Canadian coarse grain production, and more than 93 percent of grain corn production was in Ontario and Quebec with the balance in southern Alberta and Manitoba.[4]

Ontario has moved from a deficit to a surplus feed grain position, and Agriculture Canada has projected that the degree of Quebec's self-sufficiency in feed grains (now about 65 percent) will continue to increase.[5] The lessened dependency of Eastern Canada on western grains goes hand in hand with a tendency for other markets to increase in importance to Western Canadian grain producers. The export market has always been the dominant market for wheat. Export markets for Canadian barley have increased in importance since the mid-1970s. Local markets for feed grains (involving mainly on-farm use or sale to other farmers) now constitute the second most important use category for western barley. The local market has become the major market for western feed oats.

Future prospects to 1990 for the Canadian feed milling industry — and indeed, prospects for all domestic feed grain used in Canada until 1990 — will depend largely on trends and prospects for the various livestock sectors over this time period.

Livestock production, particularly cattle production in both Canada and the United States, tends to involve more intensive use of feed grains than in many other nations with well developed livestock sectors where forage consumption is emphasized to a greater extent. The increased levels of grain prices since 1973 have encouraged a reduction in the emphasis on grain feeding in Canadian cattle production relative to the late 1960s. This reduction in grain feeding, together with the gradual increases in efficiency of feed conversion into livestock products, have contributed to the tendency for domestic feed grain use in Canada to remain relatively static over the 1970s. As the subsequent discussion suggests, only a modest expansion in domestic feed grain consumption arising from increases in per capita consumption of livestock products is likely. The prospects for major effects on feed grain consumption based on increased exports of livestock products are not strong, though there is a potential for further growth in exports of pork and some possibility of growth in exports of beef.

Canadian estimates of income elasticities of demand for major meats are positive but less than one. The estimated income elasticity of demand for beef has generally exceeded that for other major meats.[6] Beef is the most frequently consumed meat in Canada, and the extent of increases in per capita consumption of this product until 1976 (when consumption peaked at more than fifty-one kilograms per person) suggested continual gradual changes in Canadian consumers' preferences towards beef. Forecasters initially viewed the subsequent lower levels of per capita consumption of beef as temporary price- or recession-induced aberrations. However, annual per capita consumption of beef has remained very stable (at close to forty kilograms per head) from 1979 to 1982. Recovery from recession conditions will likely stimulate increased per capita consumption somewhat but the previous strong preference for beef no longer seems as evident.

Per capita consumption of pork in Canada tends to vary inversely with relative prices of pork during the hog cycle. Since 1970 consumption per person has varied between a low of twenty-four kilograms (1975) and a high of thirty-two kilograms (1980).

Consumption per head of poultry meat in Canada, as in many other high income countries, increased fairly rapidly during the 1960s when technological changes in poultry production increased the supply and decreased the relative price of this product. However, during the 1970s the marketed output and price levels for poultry products became rigidly controlled by supply-management programs and poultry meat consumption per person largely levelled off, to about twenty kilograms from 1971 to 1976 and twenty-three kilograms from 1979 to 1982. Consumption of other meats (veal, lamb and mutton) is relatively minor. Small but gradually increasing quantities of fish are being consumed; per capita consumption was seven kilograms in 1981.

While the relative "mix" of domestic consumption of the various meats may change slightly in the future, the evidence of the last five years does not suggest that there will be major increases in per capita consumption of meat and most other livestock products in Canada. Economic growth may generate very modest increases in per capita consumption of some livestock products. Increasing population will also contribute only modestly to growth in aggregate consumption of livestock products over the balance of the 1980s.

The prospects for major increases in exports of livestock products from Canada are not strong. However, there could be some modest growth in exports. Canadian trade in beef and beef cattle is mainly with the United States. Both countries are characterized by high quality but relatively high-cost beef production compared to the pasture-fed beef from such major exporting countries as Australia, New Zealand and Argentina. Canadian trade in beef has been approximately balanced (we tend to be a small net importer of dressed beef and a somewhat larger net exporter of cattle). Canada does, however, have a positive balance of trade in both dressed beef and beef animals with the United States and there appears to be some potential for expansion of exports to the northwestern United States. The extent of potential increases in such exports of dressed beef are limited by the provisions of the Meat Import Law of the United States, which provides for quantitative limitations on beef exports to that nation. However, exports of live animals to the United States market are not affected by this legislation.

Canada is involved only to a limited extent in the exportation of dairy products. The most important dairy product exported, skim milk powder, is heavily subsidized. This situation is unlikely to change to any major degree. The relatively high-cost and supply-controlled poultry product sector is also unlikely to expand significantly through export markets.

Exports of pork have increased appreciably in recent years, more than tripling in value between 1978 and 1982. Over the same time period, the value of pork imports into Canada has declined. In recent years, most of the increase in Canadian pork exports has been from Eastern Canada rather than Western Canada. In 1982 the Canadian net trade balance in fresh and frozen pork was $445 million, which amounted to 93 percent of the Canadian net trade balance for meats in that year.[7] Canadian pork exports are shipped to the United States and Japan; live hogs are also exported to the United States. Canadian pork exports could well continue to increase but at a slower rate than in the early 1980s. (The outbreaks of foot-and-mouth disease in Denmark, a major pork exporter, in 1981 and 1982 reduced the competition facing Canadian pork exports to Japan. The ban on export of Danish pork to Japan was lifted in September 1983.) Nonetheless, export markets for pork offer some potential for future growth in livestock product exports, although the base is relatively small.

Feed Grain Policy and Regulation

Government policy towards the domestic feed grain sector has involved a high degree of government intervention and considerable controversy. This is probably not surprising since the interests of grain growers, livestock feeders and regions of the country frequently differ. The Canadian Wheat Board's powers to regulate interprovincial and export sales of prairie wheat were extended to prairie barley and oats in 1949. The CWB continued as the sole legal seller of these prairie feed grains across provincial borders until changes in government policy for domestic feed grains were implemented in 1974.

These changes, together with further modifications in 1976, established a dual "open market" board system for prairie barley, oats and wheat destined for livestock feeding in Canada. The board retained sole authority for feed grain export sales, although grain trading companies, including private and cooperative grain companies, became involved in the open or off-board market for these products. Trading in feed wheat, feed barley and feed oats contracts on the Winnipeg Commodity Exchange was instituted to facilitate open market pricing. The CWB continued to sell western feed grains to domestic purchasers thus becoming, in essence, a residual supplier to this market. However, the extent to which these feed grains have been supplied to users by grain companies as opposed to the CWB has varied considerably, reflecting the complex interrelationships between prices, delivery opportunities and other sources of supply. Specifically, CWB sales to the domestic feed grain market accounted for 17 percent of estimated sales through commercial channels in 1977-78. Canadian Wheat Board domestic sales fell to only 5 percent of commercial domestic sales in 1978-79, increased to 18 percent in the two following years, and accounted for a much larger proportion (49 percent) of the smaller volume of feed grains sold for domestic use in 1981-82.[8]

The CWB is required to sell feed grains for domestic use at administered price levels that reflect the relative nutritive components and market prices of United States corn and soybeans. This procedure has been criticized in Western Canada, particularly because the administered prices determined by the corn competitive formulae have from time to time been lower than the prices for feed grains in export markets — world market prices for corn, barley and feed wheat do not follow a completely fixed relationship. Government reimbursement of consequent CWB losses in export revenue was introduced in 1982.[9] CWB delivery quotas on producers' deliveries of grain sold through the off-board market were introduced in 1979 because of the congestion of grain handling facilities.[10] The Grain Transportation Authority, a government body established in 1980, has the overall responsibility of allocating grain cars for off-board grains and other western grains.

Another major component of government feed grain policy involves freight subsidies (or feed freight assistance) applied to the movement of domestic feed grains to grain deficit areas — specifically, British Columbia, parts of Northern Ontario, Eastern Quebec and the Atlantic provinces, as well as the Yukon and Northwest Territories. This program cost about $15 million in both 1981 and 1982.[11] Feed freight assistance was introduced in 1941 as a government incentive to encourage livestock production in British Columbia and Eastern Canada. The extent of program coverage was reduced in 1976. The Livestock Feed Board of Canada, a government organization that represents the interests of livestock producers outside the prairies in the area of feed grain availability, administers the feed freight assistance program. This body, together with Agriculture Canada, also administers government programs to expand the amount of feed stored in eastern British Columbia and increase inland elevator capacity in Eastern Canada. In addition, the Livestock Feed Board and the Canadian Wheat Board monitor and administer the corn competitive price formulae.

The Crow Rates: Effects on Livestock and Processing

The Crowsnest Pass rates have been another statutory program affecting the use and processing of western grains. (These rates, which have applied to rail movement of grain and designated products from the prairies to export positions, are more fully discussed in Chapter 8.) The impact of the Crowsnest rates on the volume and regional location of grain processing activities, such as livestock feeding and oilseed processing, has been the subject of considerable controversy and debate. The crops covered by the rates changed somewhat over time, and they eventually included wheat, barley, oats, rye, rapeseed and flaxseed, while designated products included flour, malt, rolled oats and pot and pearl barley. Rapeseed and linseed oil cake and oil cake meal were included on movements to Thunder Bay only and other oilseed products were excluded.[12] A government program partly offset the difference between the freight rates on oil and meal and the statutory rate for raw rapeseed and meal shipments to Thunder Bay.[13] The majority of the prairie grains shipped at the statutory rates has moved into export markets. However, the statutory rates have also applied to designated grains and products transported to export points but subsequently consumed domestically.

The statutory rates, which by the early 1980s covered less than one-fifth of the full costs of shipping grain, undoubtedly have tended, all other things being equal, to promote production of the designated grain and oilseed crops in the prairies. Furthermore, they have tended to encourage the rail shipment of these crops to export position in their raw state, instead of being processed into products not covered by the statutory rates. Livestock feeding and slaughtering are, in locational theory terms, "weight losing" activities and thus, other things being equal, tend to be attracted to areas with large feed grain supplies. The influences of statutory rates, feed freight assistance and other regional programs at least partially offset this locational pull of livestock feeding and processing to the prairie region and encouraged these activities in other regions. The influence of these programs on livestock feeding and processing has been the subject of considerable discontent for many western livestock producers.

While the general impacts of these various price distorting influences on the total volume and regional distribution of agricultural processing activity can be delineated with considerable confidence, their magnitude is much less clear. In practical terms, it is extremely difficult to quantitatively isolate these distortions from other influences, such as regional incentives to promote grain production and, more particularly, provincial incentives to encourage livestock production. Studies of the impact of the Crowsnest rates have necessarily invoked simplifying assumptions.[14] Nonetheless, the consequences of the Crowsnest rates on both the distribution of income and the allocation of resources within prairie agriculture appear to have been substantial. For example, a study by Harvey calculated that upward adjustment of rail freight rates for grain to cover long-run variable costs of shipping grain would have led to income losses to grain producers, expressed in present value terms, of $338 million and income gains to livestock producers of $234 million.[15] Harvey concluded that beef and pork production in Western Canada would expand appreciably (by 12 and 14 percent, respectively) if the Crow rates were removed, but that the effects on Eastern Canadian livestock production would be minor. This study also concluded that there would be significant positive effects in Western Canada on such sectors as meat processing, meat transport, rapeseed processing and feed processing. Though the results of this and other studies on the Crow rates indicate the potential for economic gains from adopting a system of grain freight rates that reflects the costs of shipping grain, the intersectoral and interregional distributional impacts are significant and have prevented consensus on this issue being achieved within the prairies.

The changes in the rail rates for grain resulting from Bill C-155 and from any future changes in this legislation will influence the extent to which these distortions are reduced. The changes include an expanded list of specified grains and products to which the revised statutory rates will apply (including canola and linseed oil and meal). They also involve the direct payment of the full amount of the Crow subsidy to the railways, at least until 1985-86, a feature which will reduce the incentive for adjustments in regional livestock production and related processing activities. There is further discussion on this issue in Chapter 8.

Other Grain Processing Industries

Flour Milling and Breakfast Cereal Industries

In recent years, some 13 percent of Canadian wheat production has been processed by the domestic flour milling industry.[16] Flour millers process mainly hard red spring wheat but they also mill small amounts of durum wheat (used for pasta products), Ontario soft winter wheat, and Alberta soft white spring wheat (used primarily in cake mixes, breakfast cereals and pastry flour). Minor amounts of red winter wheat, oats, rye and barley are also milled. More than two-thirds of the wheat flour milled in Canada is consumed domestically, the balance is exported. Millers have been protected from the full effect of variations in international wheat prices by the "two price wheat program." While this program has been changed periodically since its introduction in 1967, it essentially stipulates a price range within which millers pay the actual export price. When export prices have been above or below the maximum or minimum of the specified price range, millers have paid these maximum or minimum prices. Compensating payments from the federal government were discontinued in 1978. A calculation of the distribution of income transfers from this program from 1967 to 1980 indicates that, overall, there has been relatively little effect on the income levels of producers during this period.[17] In fact, the program has tended to transfer income from taxpayers to consumers.

Statistics Canada groups the flour milling and breakfast cereals manufacturing industries together since a relatively small number of firms are involved in these two industries and some flour milling firms also manufacture breakfast cereals (information on employment, shipments and value-added is given in Table 5-3). In 1981 there were 50 manufacturing establishments with 5,214 employees in the two industries. Some two-thirds of these manufacturing establishments and three-quarters of the manufacturing activity are located in Ontario and Quebec.[18] The outputs of the combined industries include flour, prepared flour mixes, breakfast cereals (including oatmeal and rolled oats) and other minor food products such as pot and pearl barley. Millfeeds, a by-product of flour milling processing, together with processed animal feeds, account for some 10 percent of the value of shipments from the combined industries. About two-thirds of the millfeeds are used domestically in animal feed. The balance are exported.

The prospects for major increases in the output of the flour milling industry are not strong. The volume of output from Canadian flour mills was relatively static over the 1960s and 1970s. During this period, flour exports from Canada tended to follow a gradually decreasing trend. Domestic consumption has tended to gradually increase, which may be attributable to gradual increases in the size of the population rather than to increases in per capita consumption. In fact, per capita consumption of flour in Canada has remained very stable, which is not surprising given that estimates of income elasticity of demand for flour are close to zero. Prospects for expansion of

Canadian flour exports during the remainder of the 1980s are not strong. Indeed, Canadian millers and exporters may be hard pressed to maintain current export volumes because of the extent of export subsidization among major exporters, particularly the EEC, and the expansion in milling activity by former flour importing regions. The tendency for only moderate expansion in aggregate domestic consumption of both flour and breakfast cereals — at a rate that reflects population expansion — will likely continue throughout the 1980s.

The Malting Industry

Some 6 percent of all the barley produced in Canada in 1980-81 was processed by Canadian maltsters. Slightly more than half of this malt was exported. The balance was used domestically, mainly by brewers and, to a lesser extent, by distillers and in other food uses. While most barley is exported from Canada as unprocessed or bulk grain, exports of barley malt have increased substantially, nearly doubling in volume over the 1970s.[19] Canadian malt exports accounted for some 8 percent of world trade in malt in 1981.[20] Malting barley is also exported as grain, although exports of barley in the form of malt have exceeded unprocessed malting barley exports since 1978-79.[21] There were three malting companies, operating eight plants, in Canada in 1982. In view of the small number of companies, Statistics Canada data on this industry are not available but are included in the miscellaneous food processing category. Other grain materials are used in the distilling and brewing industries — particularly corn, as well as lesser amounts of rye and small amounts of hops. Other industrial uses for corn include the manufacturing of high fructose corn sweetener.

Malting barley exports will likely continue to increase throughout the 1980s. The more moderate but nonetheless appreciable increase in domestic use of malt and other grain products in the brewing and distilling industries can also be expected to continue to 1990.

Oilseed Processing and Refining

The successful development of rapeseed/canola as the major Canadian oilseed crop has enabled not only substantial export of unprocessed rapeseed but has also provided the basis for most of the considerable expansion in the domestic oilseed crushing industry which occurred over the 1960s and 1970s. Import substitution of oilseeds and semi-processed and refined vegetable oil and meal products has proceeded to a considerable extent. While most Canadian oilseed exports consist of unprocessed seed, an increasing proportion are in semi-processed or processed form. (Table 6-6 shows the substantial increase in production and domestic crushing — the initial processing stage — of rapeseed and other oilseeds over the past decade.)

Virtually all rapeseed production and much of the initial domestic processing of this oilseed occurs in the prairies. The development by Canadian plant breeders of canola varieties of rapeseed (with both low erucic acid and glucosinolates content) has led to increasing acceptance and use of rapeseed oil and meal in domestic and export markets — although regulatory restrictions in the United States, which are currently being appealed, still prevent food uses of rapeseed oil in that nation. The relative importance of export markets for the major oilseed products of oil and meal is indicated in Table 6-4. By the early 1980s, more than 40 percent of rapeseed oil produced in Canada was sold in export markets, the balance being used in the domestic market. Some 60 to 70 percent of rapeseed meal was used domestically in the early 1980s, the remainder being exported.[22] In 1982 about half of the Canadian production of refined oil was derived from rapeseed which, in that year, accounted for 41 percent of Canadian-produced margarine oil, 48 percent of vegetable shortening oil and 69 percent of vegetable salad oil.[23] As Table 6-4 indicates, considerable quantities of soybeans, the world's major vegetable oilseed, are also processed by the Canadian crushing industry. While much of the domestic crushing of soybeans is based on soybean imports from the United States, the increased production of this crop in Ontario has accounted for most of the rise in domestic soybean crushing activity in recent years.

Vegetable oilseed processing involves at least two major stages. First, crushing by vegetable oil mills yields crude oil and meal. (Statistics relating to the vegetable oil milling industry are given in Table 5-2.) In 1981 there were 11 vegetable oil milling establishments with some 1,525 employees. Eight of these crushing plants were located in the prairie region. The second stage of processing involves refining the crude oil produced by these mills. This yields clear deoderized or edible vegetable oils. These refined oils are packaged, blended or further processed into salad oil, margarine, shortening and other products. Several of the oilseed crushing firms are also involved in refining operations. The other refiners, which are located mainly in Ontario and Quebec, also process imported crude vegetable oil as well as animal fats and oils. In addition, non-edible vegetable oil, in particular linseed oil from flaxseed, has a variety of industrial uses. In 1982 some thirty-one establishments in Canada were involved in refining vegetable oils.[24]

TABLE 6-6
PRODUCTION AND DOMESTIC CRUSHINGS OF OILSEEDS IN CANADA
(tonnes)

	Rapeseed/Canola		*Soybeans*		*Flaxseed*		*Sunflowerseed*	
Year	*Production*	*Crushings*	*Production*	*Crushings*	*Production*	*Crushings*	*Production*	*Crushings*
1972	1,317,700	299,581	374,800	613,719	447,500	81,015	77,100	25,027
1973	1,223,800	386,068	396,500	557,957	492,800	40,090	41,200	26,310
1974	1,163,500	292,556	300,700	716,252	350,500	N/A	11,800	15,959
1975	1,839,300	312,132	366,800	665,203	444,500	N/A	29,900	6,434
1976	836,900	382,887	250,400	695,269	276,900	N/A	24,000	N/A
1977	1,973,100	610,312	580,000	689,835	652,800	N/A	81,000	N/A
1978	3,497,200	658,235	515,600	735,490	571,500	N/A	120,200	N/A
1979	3,411,100	769,154	670,900	791,375	815,400	N/A	217,800	N/A
1980	2,483,400	940,261	713,200	1,010,788	464,800	N/A	166,100	N/A
1981	1,836,700	1,059,189	607,000	866,563	468,000	N/A	174,800	N/A
1982[1]	2,073,000	N/A	833,000	N/A	714,000	N/A	N/A	N/A

N/A = not available.
Note: [1]Preliminary.

Source: Canada Grains Council, *Canadian Grains Industry Statistical Handbook 1982,* Tables 3, 6, 20.

There has been substantial excess capacity in the oilseed crushing sector, and unused capacity increased considerably in the early 1980s. To some degree excess capacity has been encouraged by government incentives to promote regional development.[25] Members of this industry, particularly the Western Canadian rapeseed crushers, have suffered major financial pressures in recent years. Lower prices, particularly for vegetable oil, reflecting depressed world economic conditions and ample world supplies of vegetable oil, coupled with pressures to export raw seed, contributed to financial losses in 1981-82. The high prices of raw seed relative to oil and meal resulted in very low — and at times, even negative — crushers' price margins. Western crushers have argued that anomalies in the Crowsnest freight rate structure for rapeseed products contributed to their economic difficulties.

The prospects for further expansion of economic activity in the vegetable oil milling and refining industries over the 1980s can be characterized as one of moderate optimism. In general, Canadian per capita consumption of vegetable oils increased over the 1960s and 1970s. This increase was particularly evident for shortening and salad oils, while per capita consumption of animal fats — including butter and lard — decreased. These trends could moderate but are not likely to reverse to any significant degree. Plant breeders' achievements have enabled the import substitution of domestic rapeseed/canola for imported vegetable oilseeds to proceed to an appreciable extent. However, major expansions in domestic crushing activity would only be likely if major increases in export volumes of vegetable oil and meal were achieved.

Even if major growth in export markets for oilseed products takes place, public investment to encourage further crushing capacity expansion should not be entertained because of the extent of over-capacity in existing establishments. While a continuation of the trend towards increased exports of rapeseed/canola oil and meal is likely, any major expansion will be constrained by national policies of some other nations. The refusal, at least to date, by the United States Food and Drug Administration to permit U.S. generally-regarded-as-safe (GRAS) status for rapeseed/canola oil limits possible market opportunities in that nation and in other regions that tend to follow its food safety regulations. Japan, the dominant importer of Canadian rapeseed, favours the importation of raw seed rather than oil and meal. Efforts to encourage diversification of export markets are desirable, as are continued efforts to increase oil and meal exports.

Grain Handling and Transportation

Data on employment and value-added in the grain handling and transportation system are not readily available. However, some indications of the effect of the economic spin-offs from investment likely to be generated by anticipated expansion in the grain handling and transportation sectors are available. Increased investment to expand the physical capacity of the Canadian grain handling and transportation sector is anticipated to have appreciable effects on employment and income in Canada over the balance of the 1980s. Major increases in shipments through the mountains to the west coast are expected for bulk products, particularly coal, but also for potash and sulphur as well as for grain, and these will necessitate increased capital expenditures.

Canadian Pacific Railway and Canadian National Railways have started or announced major capital investments for tunnelling, grade reduction and extensive double tracking. The railways have linked further investment — particularly Canadian Pacific's Rogers Pass tunnelling project — to changes in the statutory rates for grain. The anticipated investment by the railways involves some $6.7 billion in upgrading and expansion of the rail system by 1990 (apart from maintenance of the current rail system) to accommodate projected increases in rail freight movement by 1990 — though only a portion of this investment can be attributed to increased grain shipments. Based on the relative percentages of the anticipated increases in rail traffic for different commodities, the Canada Grains Council has estimated that some 13.2 percent (or $855.5 million) of the railways' anticipated increased capital investment requirements to 1990 can be attributed to increased grain shipments.[26]

The Grain Transportation Authority anticipates that rail car requirements to meet projected increases in rail shipment of grain will involve adding an additional 7,790 hopper cars to the grain car fleet between 1983-84 and 1991-92.[27] The impact on the steel production and processing sectors in Ontario and Quebec may be gauged by the expenditure — $80 million — on the federal government's 1982 purchase of 1,280 new 100-ton capacity steel hopper cars; it has been stated that this purchase created over 550 man-years of work by the steel manufacturing firms that fulfilled the order.[28] The federal government later announced that purchases of the same number of hopper cars would be made in each of the three subsequent years.

Increased capital investment of some $300 million is anticipated in construction of a new grain terminal at Prince Rupert. This terminal, to be financed by the government of Alberta and a consortium of major Canadian grain handling companies, is expected to commence operations in 1984-85 and will add appreciably to terminal capacity at the west coast ports. Considerable expenditure on the continued maintenance and gradual upgrading of the various other components of the grain elevator system is also anticipated.

Input Supply Industries: The Backward Linkages 7

Increasing grain production has some potential impacts on the farm input supply industries. An assessment of the strength of these impacts is the issue of this chapter. Three major input supply sectors are considered: farm machinery, pesticides and fertilizers. Both the farm machinery and, in particular, the pesticide sectors have a sizeable import content (partly, but not fully, offset by a considerable export orientation in the case of farm machinery). The question of whether potential import substitution possibilities are likely to contribute to increased economic activity in these sectors is one of the issues explored in this chapter.

Agricultural Machinery

Canadian manufacturers of farm machinery are a minor part of the relatively concentrated North American farm machinery manufacturing industry. The North American industry consists of a small number of large multinational firms which generally manufacture a full line of major farm machinery (tractors, combines and a wide variety of attachments). A small number of long line producers, which are smaller and more specialized than full-line firms, produce a variety of implements, but not all the equipment, for a particular submarket. There are also numbers of relatively smaller manufacturers which produce specialized farm machinery items. These are often oriented toward particular crops or livestock, or particular conditions of the region in which they operate. Smaller companies sell their products under their own brand names or to full-line companies.

The two dominant companies in the North American farm machinery industry are John Deere Ltd. (the price leader for a majority of farm equipment items for many years), and International Harvester Ltd. The other major companies include J.I. Case (a subsidiary of Tenneco Inc.), White Farm Equipment Ltd., Allis Chalmers Corp., and Massey Ferguson Ltd. All of these firms have at some time been full-line equipment manufacturers although several of them are now tending to specialize in one or more segments of the market in response to adverse economic conditions. One of the North American "big seven" farm machinery firms — Massey Ferguson Ltd. — is Canadian based, although many of its employees are located outside Canada. This company, together with a number of other farm machinery firms (including White Farm Equipment Ltd. and White Farm Equipment, Canada Ltd.), have suffered severe financial problems in the past four years.

The most important of the smaller (long- or short-line) North American farm machinery manufacturers are Steiger (an affiliate of International Harvester), Versatile (based in Vancouver), New Idea (a subsidiary of AVCO Corporation), and New Holland (a subsidiary of Sperry-Rand).

The demand for farm machinery, while tending to gradually increase over time, fluctuates seasonally as well as changing erratically from year to year, reflecting fluctuations in farm prosperity. The larger multinational firms have been more able to weather these fluctuations because of their diversified regional markets and product lines. Barriers to entry for new or smaller firms have been substantial. Economies of scale, particularly in the manufacturing of tractors and combines, but also in distribution, financing, and research, are extensive. Studies for the Royal Commission on Farm Machinery, though fifteen years old, are still considered to give a reliable indication of the scale economies for tractor production. They showed that manufacturing costs declined by 12 percent with output increases from 20,000 to 60,000 tractors and by an additional 9 percent with an output of 90,000 tractors per year.[1] As a result, a factory price that would give a plant producing 20,000 tractors each year about a 12 percent return on invested capital would yield a 33 percent return for a 60,000 unit plant and a 45 percent return for a 90,000 unit plant.[2]

The major farm machinery manufacturing firms operate a system of regional branch house distribution centres and sell their products through networks of independent but closely supervised franchised dealers. They also operate finance subsidiaries which provide credit for dealers and for farmers' purchases. Scale economies in distributing farm machinery have encouraged the trend towards fewer and larger dealerships.[3] Dealers are discouraged from handling competing products, and thus the considerable cost of establishing a distribution system is a major barrier to potential entrants. However, in the future the decreasing number of full-line companies could tend to encourage more multi-company dealerships.

Non-price competition in the form of machinery improvements as well as model options, consumer credit terms and warranty provisions is a well established facet of market behaviour for farm machinery. Since timeliness in the use and repair of increasingly complex machinery is critical to farmers, non-price competition has also emphasized availability of repair parts and servicing. John Deere, in particular, has established and maintained a reputation for quality and reliability. This leading farm machinery company has been less affected by the adverse economic circumstances that have plagued some of the major companies in this industry since the late 1970s and early 1980s.

Despite the substantial barriers to entry in the North American farm machinery industry, some firms have been able to enter and survive or grow under the pricing umbrella of the majors, particularly when they have catered to a specialized segment of the market or introduced innovations. For example, the Canadian-based company, Versatile, has gradually moved into large tractor and combine production and concentrates on a limited number of high volume machinery items designed for the grain farming areas of the prairie provinces and the mid-western United States. This firm also manufactures specialized dryland tillage implements and plans to introduce a line of smaller tractors.

As the Royal Commission on Farm Machinery pointed out in 1971, the high entry barriers in the North American farm machinery industry, together with high prices, had fostered situations in which relatively high costs and inefficiency were more prevalent than high profits. Major firms were not fully using scale economies; emphasis on many models, sizes and options fragmented production and added to production costs; and some small plants with outdated technology continued to survive.[4] The existence of these various inefficiencies contributed to the difficulties which some of the large farm machinery firms have had in adjusting to adverse economic pressures during the late 1970s and early 1980s. Other contributing factors include the high debt that some companies incurred in expanding operations in response to the surge in demand which accompanied high farm incomes in the early to mid-1970s. High interest rates in the mid- to late 1970s exacerbated the costs of servicing this debt and also substantially increased the costs of large inventories and "floor-planning" (major manufacturers tend to finance most of their dealers' inventories).

Finally, lower demand for many farm machinery items in the late 1970s and early 1980s contributed to more emphasis on price competition during this period.

The stock of farm machinery on farms was substantially augmented when increased farm incomes in the early to mid-1970s greatly increased farmers' purchasing power. The size and relative newness of the stock of existing farm machinery and the relatively static trend in net farm income, which has been evident over the late 1970s and early 1980s, have both contributed to the prolonged decline in demand that has faced major manufacturers since the late 1970s. All the major farm machinery firms have suffered reverses. John Deere was the only major firm to record a profit in 1982.[5] Massey Ferguson only escaped bankruptcy because of refinancing involving the governments of Canada, Ontario and Britain in 1981 and subsequent refinancing in 1983. The United States firm, White Motor Corporation, filed for reorganization under the bankruptcy code in 1980 and was split into four companies. Its Canadian division, White Farm Equipment, Canada Ltd., which had been a recipient of government guarantees and assistance, including grants for the development of a new combine,[6] has since been placed in receivership.[7]

Gradual increases in demand for farm machinery are expected to be evident again in the longer-run, although the current tendency for longer replacement cycles may continue because of increasingly expensive machinery. The tendency for continued gradual enlargement of farms (implying larger but fewer machines) and some movement towards zero tillage (implying less tractor use) may also tend to dampen long-run demand prospects somewhat. There is likely to be increased concentration in this industry and its subsections since some firms may not survive the current depressed market circumstances and many are being forced to rationalize their operations by specializing in fewer areas or equipment lines.

The Canadian farm machinery industry produces about 10 percent of total North American production. Like the entire North American industry, Canadian farm machinery manufacturing is highly concentrated among a small number of larger manufacturers — the largest 4 plants accounted for 64 percent of industry shipments in 1979.[8] There is also a much greater number of small farm implement manufacturing concerns; in 1981, there were in total 207 establishments employing 16,073 workers and the value of shipments was $1,348 million.[9] Some 60 percent of workers were in plants with 500 or more employees. These plants accounted for 66 percent of value-added by farm implement manufacturers in Canada. A number of the larger firms are subsidiaries of major American multinational farm machinery companies, although the extent of Canadian ownership and control of the book value of capital employed in this industry tended to increase through the 1970s and was 49 percent in 1978.[10] The regional distribution of manufacturing activity has changed as more farm machinery manufacturing activity has tended to locate in Manitoba and Saskatchewan; their combined share of value-added was 8 percent in 1961 and had increased to 47 percent by 1981. Over this period the share of value-added in Ontario, which is still the major manufacturing centre, fell from 87 percent to 45 percent.[11] (The extent to which this industry is located in the prairies is shown in Table 7-1.) About 70 percent of Canadian farm machinery sales are in Western Canada.[12]

The Canadian composition of farm machinery production differs somewhat from that in the United States and reflects a tendency towards international specialization. Only one firm, Versatile, manufactures tractors in Canada. A relatively large proportion of Canadian farm machinery production is harvesting equipment. Canadian tariffs on farm machinery and parts were removed in 1944,[13] and since then there has been free trade in farm machinery between Canada and the United States. A large proportion of Canadian farm

TABLE 7-1
EMPLOYMENT AND VALUE-ADDED IN THE AGRICULTURAL IMPLEMENT INDUSTRY

		Agricultural Implement Industry[1]		*Agricultural Implement Industry As a % of All Manufacturing*[2]	
	Number of Establishments	*Employment (000 workers)*	*Value-Added ($000)*	*Employment (%)*	*Value-Added (%)*
Canada	207	12.2	670,333	0.9	1.0
Total Prairies	77	4.7	341,106	4.1	5.7
Manitoba	25	2.7	226,691	6.3	12.8
Saskatchewan	30	1.4	86,671	9.1	11.0
Alberta	22	0.6	27,744	1.0	0.8

Notes: [1]1981 figures.
[2]All manufacturing based on 1980 figures.

Sources: Statistics Canada, *Manufacturing Industries of Canada: National and Provincial Areas, 1980;* and Statistics Canada, *Agricultural Implement Industry.*

machinery is exported — and even more is imported. Since 1976, 90 to 93 percent of Canadian exports have been shipped to the United States and 88 to 89 percent of Canadian imports have originated there.[14] Canada also exports farm machinery to the EEC, Australia, South Africa and Mexico, while other sources of imports are the EEC and Japan. Canada exports considerable quantities of farm machinery, particularly combines, and imports substantial quantities of tractors and parts. The net trade balance in farm implements and parts has consistently been negative and has increased, particularly since 1977. The trade deficit reached a level of $1.29 billion in 1981, more than twice the 1976 deficit. Statistics Canada trade data also show the first negative trade balance in combines for many years in 1981 (see Table 7-2 on page 84).

The pessimistic short-run outlook for farm machinery demand and the prospects of only moderate increases in demand in the medium- to longer-run do not suggest that major increased economic activity in farm machinery will have appreciable effects on the Canadian economy over the balance of this decade. Success has been most evident for those Canadian farm implement firms that specialized in particular subsectors of the diverse farm implement market. Government assistance may be best directed towards encouraging such new technology as may enable successful penetration of particular market segments. For example, an expansion of the research and testing facilities of the Prairie Agricultural Machinery Institute to place more emphasis on developmental and testing activities could assist small manufacturers with technological development as could an expansion of the agricultural engineering research capabilities of Canadian universities.

Fertilizer

Increasing use of fertilizers over the past three decades has contributed to the increased output and productivity demonstrated by Canadian agriculture. The 1981 Census of Agriculture indicated that fertilizer was applied to some 40 percent of total improved farmland in Canada in 1980, compared to 16 percent in 1971. While only 38 percent of farmers applied fertilizer in 1971, almost 60 percent applied fertilizer in 1980. Between 1966-67 and 1981-82 the compound growth rate in consumption of fertilizer mixtures and materials was 6 percent per year.[15] Continued expansion of the grains sector will involve increasing use of fertilizers and agricultural chemicals, and thus some modest expansion of economic activity in these industries can be

TABLE 7-2
CANADIAN TRADE IN AGRICULTURAL MACHINERY, 1981
(million dollars)

	Exports		*Imports*		*Net Trade*	
Type of Machinery	*U.S.*	*Total*	*U.S.*	*Total*	*U.S.*	*Total*
Soil preparation	103.7	115.1	161.8	176.2	−58.1	−61.1
Crop protection	44.0	46.3	45.3	49.7	−1.3	−3.4
Haying and harvesting[1]	315.2	345.2	502.3	517.4	−187.1	−172.2
Combines	(175.8)	(193.7)	(203.9)	(207.1)	(−28.1)	(−13.4)
Market preparation	2.3	3.4	44.6	45.1	−42.3	−41.7
Dairy, poultry equipment	5.8	11.4	45.3	48.3	−39.5	−36.9
Other machinery	107.9	118.4	154.3	169.4	−46.4	−51.0
Tractor and parts	219.7	244.9	987.6	1,167.0	−767.9	−922.1
Total	798.6	884.7	1,941.2	2,173.1	−1,142.6	−1,288.4
Total 1982		651.0		1,689.0		−1,038.0

Notes: [1]Includes combines.

Sources: Statistics Canada, *Exports, Merchandise Trade;* Statistics Canada, *Imports, Merchandise Trade;* and Agriculture Canada, *Selected Agricultural Statistics Canada and the Provinces, 1983.*

expected. However, the demand for fertilizer will continue to exhibit cyclical changes, varying, in particular, with the fortunes of the crop sector (annual fluctuations in the growth of fertilizer consumption have ranged from −16.1 to 21.5 percent from 1966-67 to 1981-82).[16]

Fertilizer production requires quantities of specific mineral inputs, some but not all of which are available in Canada. The three main fertilizer nutrients are nitrogen, phosphorus and potassium. Nitrogen is generally supplied in some form of ammonia which is produced by combining atmospheric nitrogen with hydrogen from a feedstock such as natural gas. Phosphorus is generally applied in the form of ammonium or calcium phosphates which are manufactured from mined phosphate rock and sulphuric acid. Large quantities of phosphate rock are imported into Canada from the United States for this purpose. Potassium is applied as potash fertilizers; these materials are mined and their processing mainly involves washing, concentration and mixing without chemical treatment. Large deposits of potash in Saskatchewan provide the basis for major exports of this commodity to the United States and to other regions. Some 66 percent of Canadian potash exports in 1980-81 were shipped to the United States[17] and accounted for the equivalent of 70 percent of United States potash consumption in 1980.[18]

A single trading area with north-south flows has developed in North America. The fact that tariffs and other trade barriers are virtually non-existent has fostered international trade and has reinforced the division of the Canadian fertilizer industry into Eastern Canadian and Western Canadian segments. Statistics Canada does not present separate data on the major subsector of the Canadian fertilizer industry — the primary fertilizer manufacturing sector. Data for this sector are included in those provided for the industrial chemicals industry. In 1981 there were some thirty basic fertilizer plants, including twenty plants producing nitrogenous and/or phosphatic fertilizers (Table 7-3) and ten potash mines located in Saskatchewan (Table 7-4). This primary sector had some 10,000 to 12,000 employees in 1981. A number of new fertilizer plants were brought onstream in Western Canada during the 1970s and capacity was further expanded in the early 1980s. The industry shows evidence of over-capacity, at least in the shorter-run.

As Table 7-3 indicates, the major Canadian fertilizer manufacturing companies include Cominco Ltd. (the majority of which is owned by Canadian Pacific Enterprises Ltd.), CIL Inc. (largely owned by Imperial Chemical Industries Ltd., Great Britain), Western Coopera-

TABLE 7-3
PRIMARY FERTILIZER MANUFACTURERS: AMMONIA AND PHOSPHATE FERTILIZERS, 1981 ANNUAL CAPACITY
(thousand tonnes)

Company	*Location*	*Urea*	*Ammonia*	*Ammonium Nitrate*	*Ammomium Phosphate*	*Superphosphate*	*Wet Process Phosphoric Acid*
Belledune Fertilizers (Noranda)	Belledune, N.B.	—	—	—	272	—	136
Beker Industries	Sarnia, Ont.	—	idle (145)	—	—	—	—
Canadian Fertilizers Ltd.	Medicine Hat, Alta.	435	360	—	—	—	—
Canadian Fertilizers Ltd.	Medicine Hat, Alta.	—	360	—	—	—	—
CIL Inc.	Beloeil, Que.	—	—	65	—	—	—
CIL Inc.	Courtright, Ont.	160	360	145	170	—	85
CIL Inc.	Carseland, Alta.	435	—	225	—	—	—
Cominco Ltd.	Calgary, Alta.	70	107	20	—	—	—
Cominco Ltd.	Carseland, Alta.	—	360	—	—	—	—
Cominco Ltd.	Kimberly, B.C.	—	—	—	160	—	86
Cominco Ltd.	Trail, B.C.	—	65	—	175	—	76
Cyanamid of Canada Ltd.	Welland, Ont.	91	221	200	—	—	—
Esso Chemicals	Redwater, Alta.	—	225	210	430	—	205

TABLE 7-3 (Continued)
PRIMARY FERTILIZER MANUFACTURERS: AMMONIA AND PHOSPHATE FERTILIZERS, 1981 ANNUAL CAPACITY
(thousand tonnes)

Company	*Location*	*Urea*	*Ammonia*	*Ammonium Nitrate*	*Ammonium Phosphate*	*Superphosphate*	*Wet Process Phosphoric Acid*
International Minerals & Chemical Corp.	Port Maitland, Ont.	—	—	—	50	69	118
Nitrochem Inc.	Maitland, Ont.	45	80	170	—	—	—
St. Lawrence Fertilizers (Noranda)	Valleyfield, Que.	—	—	—	idle (54)	idle (60)	idle (45)
Sherritt Gordon Mines	Fort Saskatchewan, Alta.	80	145	—	123	—	48
J.R. Simplot Co.	Brandon, Man.	27	100	135	145	—	—
Western Cooperatives	Medicine Hat, Alta.	—	60	77	181	—	67
Western Cooperatives	Calgary, Alta.	—	60	60	260	—	135
Total Capacity		1,343	2,503	1,307	1,966	69	956

Source: Canadian Fertilizer Institute, ''Amended List'', 7 October 1981.

TABLE 7-4
PRIMARY FERTILIZER MANUFACTURERS: POTASH, 1981

Company	*Location*	*Annual Capacity (thousand tonnes)*
Central Canada Potash (Noranda Mines 51%, CF Industries 49%)	Colonsay, Sask.	817
Cominco (Canadian Pacific Enterprises)	Vade, Sask.	653
International Minerals & Chemicals Corp.	Esterhazy, Sask.	1,161
International Minerals & Chemicals Corp.	Esterhazy, Sask.	953
Kalium Chemicals (PPG)	Belle Plain, Sask.	850
Potash Corp. of America (Ideal Basic)	Saskatoon, Sask.	421
Potash Corp. of Saskatchewan	Allen, Sask.	515
Potash Corp. of Saskatchewan	Cory, Sask.	649
Potash Corp. of Saskatchewan	Esterhazy, Sask.[1]	581
Potash Corp. of Saskatchewan	Lanigan, Sask.	623
Potash Corp. of Saskatchewan	Rocanville, Sask.	622
Total Capacity		7,845

Notes: [1]Mining and refining contract with I.M.C.

Source: Canadian Fertilizer Institute, "Amended List," 7 October 1981; and R.F. Liebenluft, *Competition in Farm Inputs: An Examination of Four Industries*.

tive Fertilizers Ltd., Imperial Oil Ltd., Sherritt Gordon Mines, Ltd., and Simplot Chemical Company Ltd. Ownership of the ten potash mines and related refineries was originally dominated by major American potash mining companies, but legislation passed in 1976 enabled the government of Saskatchewan to purchase control of an appreciable proportion of the industry. International Minerals and Chemicals, the Potash Corporation of Saskatchewan, Ideal Basic Industries (Potash Corporation of America) and PPG Industries are now the four largest North American potash producers, accounting for a combined market share of 59 percent of total North American potash production.[19]

In general, the primary fertilizer manufacturing sector is characterized by a number of large, capital intensive, world-scale plants. Most fertilizer used in Canada is manufactured domestically and a considerable proportion of Canadian fertilizer output is exported. Some 76 percent of the total volume of commercial fertilizer shipments by primary manufacturers was sent to export markets in 1981-82.[20] (A large proportion of total exports is composed of muriate of potash.) Canada has a substantial trade surplus in fertilizer; this surplus grew through the 1970s, reached $1.18 billion in 1980-81, but declined slightly to $1.03 billion in 1981-82 because of weaker world markets.[21]

Most of the primary producing firms control or own blending facilities or bulk storage facilities. In Western Canada, retail distribution largely occurs through local agents of the manufacturers and farmers' cooperatives. In Eastern Canada most fertilizer is sold to farmers in the form of mixed blends and the distribution system involves a number of smaller blending plants.[22] Manufacturers of mixed fertilizers, which mix and blend major fertilizer components, operated 105 plants in 1981 and employed some 1,000 workers. Most of these manufacturers are found in Ontario and Quebec. There are, in addition, fifteen establishments that manufacture lime in Canada, also located primarily in Eastern Canada.

One study by Rennie and co-workers suggested that nitrogen fertilizer use in Western Canada might triple between 1980 and 1990.[23] This estimate, based on an assumed 50 percent decline in summer fallow and a decline in soil nitrogen availability, might be somewhat high. However, the compound growth rate of nitrogen consumption in commercial fertilizers in Western Canada rose from 8.8 percent per year from the mid-1960s to the early 1970s to 11 percent from the mid-1970s to the early 1980s.[24] Fertilizer production requires a relatively small proportion of the natural gas produced in Alberta and hence availability of this feedstock is not expected to constrain continued expansion of nitrogenous fertilizer production.[25] Rennie also forecast that phosphate consumption would increase by somewhat more than half between 1980 and 1990 (consumption of phosphate in commercial fertilizers increased by an annual compound rate of 5.1 percent in Western Canada from the mid-1960s to the early 1970s and this rate fell to 3.3 percent from the mid-1970s to the early 1980s[26]). Very small amounts of potassium-based fertilizers are applied in Western Canada, although consumption of these materials has increased and is projected to continue to rise, particularly in irrigated and high-producing areas.[27]

The increases in fertilizer use resulting from continued expansion of the agricultural sector have been anticipated, to some degree, by major expansions in fertilizer manufacturing capacity in recent years, particularly for nitrogen but also for phosphatic fertilizers. However, continued moderate expansion in the grains sector in the medium- to longer-run will likely generate some further growth in this industry.

Agricultural Chemicals

Agricultural chemicals, or farm pesticides, include fungicides to control seed decay and seedling diseases, insecticides to control crop and livestock insect pests, and herbicides to control weeds. The use, quality and variety of these products have increased greatly over the past three decades. Canadian farmers' expenditures on such products amounted to $420 million in 1982.[28]

Herbicides account for the largest proportion of agricultural chemical sales in both Canada and the western provinces; herbicide sales exceeded 70 percent of Canadian agricultural chemical sales in 1981.[29] Wild oats are currently considered the major weed problem for western cereal crops, and herbicides applied for wild oat control were estimated to account for more than 70 percent of the value of herbicides applied in Western Canada in 1978.[30] Broad-leaf weeds are also a problem, and some 60 to 70 percent of cereal crops are sprayed annually with 2,4-D and MCPA.[31] Insecticide sales tend to vary in response to major pest outbreaks. Use of fungicides is relatively smaller but may rise with increasing acreage devoted to rapeseed.[32]

The increasing use of farm chemicals since the 1940s has enabled substantial growth in agricultural output. However, chemical control of pests has sometimes led to adverse effects on both the environment and people and is thus the focus of some public attention and concern. Consequently, regulations have increasingly governed the development and use of farm chemicals. New chemicals must be approved and licensed by Agriculture Canada, while agricultural use of chemicals that leave persistent residues has been prohibited or limited.

Cultural methods, plant breeding for disease or pest resistance and biological control are also important facets of pest control. Nonetheless, chemical control of agricultural pests is and will continue to be an important aspect of Canadian agriculture. While demand for agricultural chemicals has moderated somewhat in recent years, this industry has been less adversely affected by recessionary conditions than suppliers of other major agricultural inputs. In Western Canada, the tendencies for less summer fallowing, longer crop rotations, more emphasis on continuous cropping and less tillage in crop production have been fostered by the recognition that the extent of summer fallow practices may have contributed to soil deterioration. However, these tendencies are expected to lead to more extensive weed problems and perhaps some increase in disease and insect problems. Increased use of agricultural chemicals, in particular more specific and effective chemicals, can be expected in the future.

The Canadian agricultural chemical manufacturing industry is essentially an industry that formulates imported active ingredients. These active chemicals are imported mainly from the United States but also from the United Kingdom, Europe and Japan.[33] Active (or basic) ingredients are based on petrochemical feedstocks, and their production involves a high capital investment and technical capacity in research programs and process facilities. Studies of the active chemical production sector in the United States indicate that industry concentration levels tend to be high in individual submarkets (for crops or pests). However, the use of leading products can change substantially in a short time as a result of pest resistance, product suspension, patent expiry or the development of improved products.[34] Product development costs have increased markedly in recent years since regulatory requirements in the United States have increased testing costs of firms and extended the lead time before a return can be earned on a successful new product. It has been estimated that the cost of developing a new agricultural chemical rose from about $3.4 million (U.S.) in 1967 to approximately $20 million in 1980.[35] Most manufacturers of active ingredients are large companies which process a range of industrial chemicals, although some also manufacture pharmaceutical products.

In 1979 the sole Canadian producers of active agricultural chemical materials were two subsidiaries of American industrial chemical firms. The balance of active ingredients used in Canada — 96 percent in that year — was imported. The two plants in Alberta involved in basic production were not world-scale plants but produced relatively small amounts of agricultural chemicals, mainly 2,4-D esters.[36] However, the use of 2,4-D esters on crops is being replaced by 2,4-D amines and MCPA which cause less plant damage. By 1983 there was virtually no synthesis of active ingredients in Canada and the Canadian agricultural chemical industry was one of formulation.

Agricultural chemical formulation involves mixing active ingredients with emulsifiers, gypsum and solvents to stabilize the chemicals for marketing. Basic pesticide producers are performing these functions to a greater extent, and thus large and increasing volumes of Canadian agricultural chemicals are imported into Canada in ready-to-use form. In 1975 some 50 percent of ready-to-use product was imported.[37] The proportion of imports comprised of ready-to-use products increased through the late 1970s.[38] In 1981 this proportion was reported as nearly 80 percent.[39]

There are some 35 to 40 firms formulating agricultural chemicals in Canada; in the mid-1970s they employed about 1,200 workers.[40] Employment by Canadian formulators seems to have remained at about this level in subsequent years.[41] A small number of these establishments account for a large proportion of agricultural chemical sales; three firms accounted for 65 percent of sales in Canada in 1975.[42] A large proportion of the Canadian formulating companies are subsidiaries of foreign (mostly American) manufacturers of active ingredients.[43] Research expenditures in Canada do not appear to be high: it was estimated that for the 12 major companies carrying out research and development in 1975, 113 persons were involved and $1.5 million was spent on product development.[44]

Virtually all active ingredients and a large proportion of the formulated agricultural chemicals used in Canada are imported. In 1981-82 these imports were valued at $163 million while exports amounted to $14 million.[45] Canadian agricultural chemical companies have argued that the international structure of tariffs for agricultural chemicals has contributed to this pattern of trade. In the interests of moderating agricultural input costs for the farm sector, major agricultural production inputs, including active agricultural chemical ingredients, entered Canada duty free for many years. However, a protective tariff of 15 percent was imposed in 1977 on the importation of phenoxy herbicides (2,4-D and MCPA).[46] Although the tariff was intended to encourage the Canadian production of 2,4-D, it has not been successful. The size of the domestic market is such that a portion of output of any basic ingredient from a world-scale plant would need to be marketed outside Canada. A policy of import substitution, based on the infant-industry argument for initial tariff protection, could result in the establishment of a small-scale, inefficient and high-cost industry which could never withstand the rigours of the international market. A major barrier to entry into production of active agricultural chemicals is the research and development expenditures and expertise required for development and licensing of improved new farm chemical products.

The prospects for substantial increases in agricultural chemical production, particularly of active chemicals, is not strong. In terms of the volume and value of agricultural chemical use, the most likely candidates for further study of the possibilities for import substitution are wild oat herbicides and 2,4-D amines. The large multinational companies that dominate the Canadian agricultural chemical industry do relatively little of their research and development within Canada. How increased research and development activity directed at the major Canadian pest problems should be encouraged is a further question which should be addressed in the public policy arena.

Policy Issues and Constraints 8

Improved policy for grains and oilseeds would assist in realizing the moderate growth opportunities which are currently foreseen for this sector. In this chapter, we examine several selected policy areas which impinge critically on the grains and oilseeds sector: the Crow rates issue; grain price and income stabilization; the extent of income support for the grains sector; grain institution performance and reform; research and productivity concerns; land, water and environmental issues; and, finally, macroeconomic and trade policy.

The Crowsnest Pass Rates

In the 1970s and early 1980s controversy has surrounded the statutory rates on rail shipment of grains and grain products from the prairies to export points. The Crow rates were first instituted in 1897, subsequently altered, and then given statutory status in 1925. They were set at a rate of one-half cent per ton mile and eventually applied to wheat, barley, oats, rye, rapeseed, flaxseed and such products as flour, malt, by-products of the milling, distilling and brewing industries, and certain feed grain by-products. The rates have applied on rail shipments to the Lakehead, west coast ports and Churchill. In recent years the Crow rates have been considerably less than the costs of rail shipment of grain and thus involved an appreciable subsidy to western grain growers. Indeed, as a following section of this chapter points out, these rates have been the main form of subsidy support for western grain growers, though their benefit to producers was at least partly offset by the deterioration in railway service which they induced. The other major impact of the Crow rates was their effect on the nature of the prairie crop and production mix and the extent of grain-processing activity in the prairie region. This effect was also discussed in Chapter 6.

The Crow rates involved considerable costs to the railways, at least from the early 1960s. The shortfalls in railway costs of moving grain over the revenue received under the statutory rates were exacerbated throughout the 1970s and early 1980s by the effects of inflation on railway costs and by the increasing volumes of grain movements. Snavely concluded that in 1980 the variable costs incurred in moving statutory grain amounted to $547.5 million, while revenues from shippers of statutory grain amounted to $132.9 million.[1] While certain government expenditures compensated for part of this loss[2] (these expenditures were estimated at $170.2 million for 1980), the consequent annual shortfall in coverage of variable costs was substantial ($244.4 million) and had been increasing. Snavely noted that the attribution of an appropriate share of fixed costs to statutory grain movements could only be subjective and suggested an allocation of 22.5 percent of variable costs. This implied that estimated total railway losses (after accounting for user and government contributions) were $381.3 million on statutory grain movements in 1980.

One major effect of this situation, not surprisingly, was considerable deterioration in the physical plant and efficiency of operations and service available to shippers of statutory grain. In his report on Western grain transportation, Gilson noted that between the 1960s and 1980 the railways made virtually no investments in rehabilitating grain rolling stock or branch lines and maintenance work had been minimal.[3] The constraints of the handling and transportation system contributed to lost export sales by the Canadian Wheat Board, disruptions in deliveries, and substantial demurrage charges (the penalties paid to ship owners for delayed loadings). Demurrage charges alone on delayed loadings of wheat, oats and barley reached a peak in 1977-78 and amounted to close to $22 million.[4]

Increasing volumes of exports were achieved by a variety of measures, including the institution of block shipping and car pooling programs in the late 1960s. The federal government instituted the branch line subsidy program; subsidies were paid to the railway to retain in service those uneconomic branch lines considered necessary for grain shipment. A subsequent program was instituted to upgrade and rehabilitate those branch lines specified as part of the basic branch-line network. A cost-sharing agreement to repair old box cars was initiated. The grain car fleet was augmented by the federal government's purchase and long-term lease, over the period from 1972 to 1981, of some 10,000 grain hopper cars. Since then additional hopper cars have been purchased by the federal government. Hopper cars have also been purchased by the CWB (2,000) and the governments of Alberta (1,000), Saskatchewan (1,000) and Manitoba (400 leased). The Grain Transportation Authority, a federal government agency, began operations in 1980 in an attempt to improve coordination and planning in the shipment of increasing volumes of grain. This body undertakes major responsibility for grain car allocation for board versus non-board grains and among shippers of non-board grains.

Improvements in port and terminal capacity and the initiatives noted above, together with lessened competition from other freight (a feature of the global recession of the early 1980s), have enabled record shipments of grain in 1981-82 and 1982-83. Nonetheless, upgrading of the system is required to enable the railways to carry the increasing volumes of freight which are forecast for the balance of the 1980s, particularly for shipments through west coast ports.[5] Shippers of bulk products pressed for changes in the statutory rates for grain, and the railways linked necessary upgrading, particularly completion of Canadian National Railway's main line double-tracking and the construction of the Canadian Pacific Railway's new tunnels and bridges through Rogers Pass, to contributions from increased revenue from carriage of grain.

The federal government initiated a consultative process, chaired by J.C. Gilson, in 1982 and subsequently proposed legislation to change the statutory rates. Following extended and stormy debates, legislation to alter the system of statutory rates was passed in November 1983. The legislation, Bill C-155 or the Western Grain Transportation Act, covers an expanded list of grains and products. It involves the federal government paying the specified Crow benefit (based on the 1981-82 railway revenue shortfall, specified as $658.6 million[6] and confined to 1981-82 shipment volumes[7]) as an annual subsidy.

Major disagreements on the Crow rates have involved the method of payment of this annual subsidy and have arisen from the conflicting interests of grain growers, livestock producers and different regions. The initial proposal by the federal government involved the Crow subsidy being split evenly between railways and farmers by 1985-86. Following intensive lobbying by the three prairie wheat pools and Quebec farm groups, the method of payment was changed. Bill C-155 provides for payment of the subsidy entirely to the railways, but it also provides for a review committee to study and to report on the issue by April 1985.

Payment of the subsidy to the railways implies that rail rates for grain will continue to be much lower than the costs of rail shipment of grain. However, grain shippers will pay a share of future railway cost increases. It is anticipated that there could be a doubling of grain rail rates by 1986 and a five-fold increase by 1991. In contrast, payment of the Crow benefit subsidy to farmers would imply higher rail transport rates for the specified grain and products and, all other things unchanged, lead to a more rapid and extensive reduction of the effect of a distorted structure of freight rates on the regional location of livestock feeding and processing activities. The decision that payment of the Crow benefit will be made directly and entirely to the railways will reduce the depressing effect that otherwise would be expected to influence the farm level prices of grain in the prairies. However, this method of payment will also reduce the incentive for adjustments in regional livestock production and other grain processing activities. Specifically, it will leave unchanged those Crow-induced distortions among grain, livestock and related processing activities that had developed by the early 1980s[8], although the feature that grain farmers will pay a share of any future increases in the cost of shipping grain will reduce the extent of future increases in these distortions.

Price and Income Instability

Considerable variability in farm prices and incomes is a longstanding feature of agriculture. This variability stems partly from the effects of unplanned variations in supply, due to weather or pests, which interact with price-inelastic demand functions to accentuate price variability for farm products. The relatively long production process for some agricultural products and the difficulties in accurately forecasting their prices contribute to cyclical patterns of output and price for livestock products and perennial crops, but is less evident for prairie grains. However, variations in demand, particularly in export demand, which interact with supply schedules that are generally price-inelastic in the short-run, also accentuate price and income variability for Canadian grains and oilseeds.

The extent of price and income variability for Canadian agriculture intensified in the 1970s, compared to the 1960s. The longstanding feature of price and income variability is expected to continue. However, there is some indication that the extent of this variability was appreciably less in the latter part of the 1970s and early 1980s than in the first half of the 1970s.[9] The question of the cause of this feature and whether it is transitory or not warrants further examination.

Observers of the extent of instability in agriculture have often focused attention on market intervention programs as the major means to counteract the effects of instability. Whether such a response is appropriate depends partly on the nature and source of instability for the particular agricultural product as well as on the extent of costs of the specific stabilization programs relative to the extent of costs of the instability itself. In contrast to the situation for beef and hogs, year-to-year changes in the production of wheat, oats and barley appeared to be more major contributors to gross income variation than price changes, at least from 1946 until the early 1970s.[10] Perhaps this no longer applies, since the extent of price variability for grains in world markets has greatly widened since 1973.

Nonetheless, concern with instability should relate to the nature and source of variability in prices and incomes. Production variability is one contributor to price and income variability, and programs to reduce the extent and effects of yield variability have been and continue to be important for prairie grains. For example, the achievements of plant breeders in developing pest-resistant and earlier maturing varieties have reduced uncertainty and production variability for prairie grain and oilseeds.

The three major programs directly focused on counteracting instability in the western grains and oilseeds sector are crop insurance, the price pooling program of the Canadian Wheat Board, and the Western Grains Stabilization Program.

Crop insurance involves joint federal-provincial programs whose coverage, both in terms of crops and the proportion of participating farmers, has continued to grow since they were instituted in the early 1960s. In Western Canada, these programs eventually replaced the Prairie Farm Assistance program (PFAA). The programs are voluntary and some 52 percent of Canadian Wheat Board permit holders participated in crop insurance programs in 1978-79; the highest proportion of participating farmers (58 percent) was in Saskatchewan where specialized crop production is more prevalent than in Alberta and Manitoba.[11] The programs enable farmers to purchase crop insurance against reduced yields, due to weather and pests. The premiums and the administration of the programs are subsidized: in the prairie provinces, farmers pay 50 percent of the premiums, the other 50 percent is paid by the federal government, and the provincial governments fund the costs of administering the programs. The federal government also provides loans or reinsurance against major crop disasters, and the provincial governments operate these programs.

Since crop vulnerability to weather and pests generally increases with the intensity of production, the importance of the programs and the number of farmers enrolling in them will likely continue to increase. Federal expenditures on the crop insurance programs, which now cover all provinces and apply to a wide range of crops including forage and certain perennial and specialized crops, amounted to slightly more than $100 million in 1980-81; some 79 percent of this was paid to the three prairie provinces.[12]

Compared to other systems of price stabilization through market or price intervention, price pooling within a single production period represents a relatively low-cost and low-risk method of reducing short-run variability in prices paid to producers. Within-season price pooling has been a feature of the Canadian Wheat Board since its inception and was also a characteristic of the board's predecessors. It involves payment to producers of net realized average prices, subject to type and grade of grain and subject also to deductions for elevating and for the transport costs from the producer's delivery point to either Thunder Bay or Vancouver, regardless of delivery date during the crop year.

In practice, government-specified and government-guaranteed initial payments are set for the various grades of prairie wheat, oats and barley on the basis of delivery either at Thunder Bay or Vancouver. Producers receive this (net of primary elevator and freight costs involved in moving the grain to terminal position) at the time of delivery, whenever grain is delivered for sale by the board. Six pools are maintained by the board: wheat, durum wheat, designated barley (barley for malting and other food uses), other barley, oats and designated oats (oats for processing and milling for food uses). Sale receipts are credited to these pools, and deductions for interest, storage, insurance, terminal elevator charges, other handling charges, demurrage and the board's administration costs are debited from them. Final payments are made to producers from any balance remaining when the pool is closed — that is, when most stocks are sold and any remaining stocks are priced and transferred to the pool for the subsequent year.[13]

The federal government operates two direct stabilization programs for agriculture. These differ in a number of ways. The Western Grains Stabilization Program applies to the major prairie grains and oilseeds and is intended to prevent farmers' aggregate annual net cash receipts from falling below the average of the preceding five years. The Agricultural Stabilization Act Program applies to other major agricultural products.

The Western Grain Stabilization Program has operated since 1976.

It is a voluntary program, although there are limitations on the extent to which farmers can opt in or out. It is partly funded by participating farmers, who pay a levy of some 2 percent of their net cash receipts — to a specified, maximum annual contribution for each participant — on their sales of wheat, barley, oats, rye, rapeseed, flaxseed and mustard seed. (Some changes in the maximum level of annual contribution and in crop coverage have been made since the program started.) Farmers' levies are matched on a $2 for $1 basis by the federal government, which also bears the cost of administering the program. Contributions earn interest which is credited to the stabilization fund, and there are provisions for the levy to be altered depending on the level of the stabilization fund. The federal contribution to this fund amounted to $97 million in 1980. The aggregate net cash flow (aggregate receipts from the specified grains less specified cash costs) of participating producers is determined annually. When, for any one year, this falls below the average calculated for the preceding five years, a stabilization payment of the difference is paid to participating producers. This is shared among producers on the basis of their previous contributions to the fund.

During the first seven years of the program, from 1976 to 1982, payments were made for two years, 1976 and 1977. These amounted to $115 million for 1977 (the average payment for participating producers was $896) and $253 million for 1978 (when the average payment was $1,843). Some 76 percent of eligible producers are participants.[14]

Stabilization or Support?

The stabilization programs for western grains involve some elements of public support but have avoided the problems and economic costs associated with major subsidization or support programs. Effective agricultural stabilization programs are those programs that reduce very short-run or random price variations. They should thereby reduce, to some degree, the extent and costs of uncertainty and risk facing farmers, without distorting those price signals that would indicate more fundamental changes in resource allocation decisions — for example, those stemming from changes in the underlying medium- and longer-term trends in demand or cost. Rather than focusing on programs that are primarily intended to mitigate the extent and effects of instability, domestic agricultural policies in many nations, particularly those with relatively high incomes, have tended to support rather than stabilize. These policies have increasingly involved trade protection and price and income support measures. Such programs are particularly evident in the agricultural subsectors whose comparative advantage is weaker. In North America, for example, protection and support have been particularly evident for dairy production in both Canada and the United States, for poultry production in Canada, and for sugar production in the United States. A lesser degree of support has been extended to the grains sector, where ability to compete in international markets is evident.

Comparing the nature and extent of support to grains and oilseeds production in Canada with that in the United States, it is evident that the major support mechanism for this sector in Canada comes from subsidized transportation costs. Canadian policy for grains does not include price or income supports. (The operation LIFT program in 1970 was a temporary exception.) In contrast, support to the grains sector in the United States has mainly focused on various price and income support programs. The extent of support for the American grains sector has varied considerably, changing with economic circumstances and prevailing political philosophies. Levels of support were generally relatively high during the 1960s. However, they declined substantially in the mid-1970s but have tended to increase again in the early 1980s.

Studies by Glenn, Carter and Tangri estimated that total subsidies for wheat in the United States, expressed as a percentage of the total value of wheat production, averaged 18.5 percent over the 15-year period from 1965 to 1979. They estimated the Canadian average subsidy for wheat over that period as 11.6 percent.[15] From 1970-71 to 1980-81 the U.S. subsidy averaged 12.5 percent and was considerably more variable than the Canadian average subsidy of 14.5 percent.[16] This study also estimates support levels for American-produced barley, corn, sorghum and soybeans and for Canadian-produced barley, oats, rapeseed and flaxseed. In general, subsidy levels for barley were similar to those for wheat in Canada, while the levels of subsidy for the other grains were somewhat less. Estimated support levels for soybeans in the United States were very low.

Support for grain production has traditionally been much less in North America than in Western Europe, though this was not the case in 1973-74 and 1974-75 when greatly increased world graïn prices exceeded or were close to EEC threshold prices (the minimum prices at which grain can enter the EEC). Roberts and Tie have estimated that for 10 years, including and prior to 1981-82, the average transfer to EEC grain producers, expressed as a percentage of the value of production, was 33 percent.[17] This level of support has encouraged grain production to the point that by 1981-82 the EEC had become a net exporter of grains.

Much increased grain support levels prevailed in the United States in the early 1980s although these were still below those in the EEC. The Canadian Wheat Board has estimated American and Canadian subsidies for wheat for 1982-83; the estimated subsidy transfer per bushel in Western Canada (U.S. $0.37 per bushel) was one-third of that prevailing in the United States (U.S. $1.12 per bushel).[18]

The major component of subsidy support to the Canadian grains sector relates to the statutory rates on shipment of prairie grain to export positions. The changes in the Crow rates are likely to reduce the relative proportion of future subsidy support for Western Canadian grains. Selective strengthening of Canadian grain stabilization programs, rather than provision of alternative support programs, is a more desirable mechanism of maintaining the competitive position of Canadian grain producers relative to competitors in export markets. The expected reduction in the relative proportion of subsidy support to western grain growers must also be offset by a continued search for improvements in technology and productivity in the production, handling and transportation of grains in Canada.

Comparison of Grain Marketing Systems in the United States and Canada

A number of studies in recent years have attempted to assess the performance of North American grain marketing systems. Over the 1960s and 1970s Canadian grain exports did not increase as rapidly as world trade in grain or grain exports from the United States, and this focused attention on assessing marketing performance in Canada. Interest in the United States in the Canadian institutional structure for marketing export grain was raised by its experience with Russian purchases of unprecedented quantities of American grain in 1972. Critics of the procedures involved in those purchases called this "the great grain robbery" and argued that the USSR and a small number of large private multinational grain trading firms had benefited at the expense of American grain farmers and taxpayers.[19] The increasing number and importance of state trading agencies as purchasers and importers of grain also contributed to interest in the United States regarding the functioning and performance of its existing grain exporting sector.

Watt, Mitchell and Ross compared performance in export markets for the Canadian and American wheat marketing systems from 1971-72 to 1978-79 and concluded that the Canadian system had been more successful in selling a larger portion of its exportable supplies of wheat in each of those years except one. They also concluded that Canada was able to exert market power in export pricing during periods from 1972 to 1977 whereas the United States had not.[20] In McCalla and Schmitz's comparison of the American and Canadian grain marketing systems,[21] they suggested that the system in the United States is more efficient in many dimensions (except in efficiency of certain information flows and the merits of product quality consistency and grading), but that it is difficult to conclude that either system is superior.

Several studies have attempted to compare the pricing performance of the two systems. Foodwest Resource Consultants concluded that returns to producers in 1978 were higher under the American system than the Canadian.[22] These conclusions were challenged by the Canadian Wheat Board which contested the price comparisons used.[23] Peltier and Anderson compared prices received by farmers for wheat in North Dakota and in Manitoba and concluded that net prices to both groups of farmers were approximately the same, despite freight rates to export points being considerably lower in Canada than for comparable movements in the United States.[24] The effect of grade differences on this comparison has been raised.[25] Harvey found similar results to Peltier and Anderson.[26] The implication may be drawn that over the period studied (1970-71 to 1977-78) much of the subsidy benefits of the Crow rates to prairie grain producers were offset by constraints and costs in the Canadian grain transportation and handling system.

Overall, the Canadian system involves a higher priority for objectives of short-run price stability and relatively equal access to market opportunities and returns than the American system. Pursuit of these objectives may have involved some trade-offs with efficiency. However, attributing differences in performance of the two systems to such features as the presence or absence of a grain marketing board may be misleading. There are other differences in the institutional structures which affect each system and major differences in their physical environments. It is unlikely that the institutional structure for grain marketing of either country could be readily transplanted or accepted by the majority of grain producers in the other environment.

Research and Productivity

Productivity advance, or technical change, is generally regarded to be the mainspring of agricultural growth. Agricultural research and development, in turn, lie at the heart of productivity improvement. Research on grains and oilseeds has been very important to the crops sector in Western Canada, as is readily seen in the development of early-maturing, high-quality, and successively more rust-resistant wheats or the recent emergence of improved canola varieties. There are many concerns and doubts, however, about the current adequacy of policy for research and productivity enhancement relating to the grains sector.

An overall concern is that productivity advance in agriculture in Western Canada may have been lagging in the 1970s. In Figure 8-1, preliminary estimates of flexible-weight (Divisia-related) indices of aggregate output, aggregate input use, and total factor productivity are presented for Western Canadian agriculture (encompassing both the crops and livestock sectors) for the period 1961 to 1980.[27] From 1962 to 1980 (eliminating 1961, a bad drought year), aggregate output is estimated to have increased at the annual compound growth rate of 2.0 percent, aggregate input by 1.0 percent, and total factor productivity by 1.0 percent. Total factor productivity — the ratio of aggregate output over the aggregate of all inputs used in production — is a superior productivity measure to such commonly used partial productivity measures as labour productivity (output per man or per man-hour) or land productivity (yields per acre). Productivity measurement involves many serious conceptual and empirical problems. In addition, productivity measures for agriculture in Canada appear to be very sensitive to the particular time period under consideration. Nevertheless, as evidenced in Figure 8-1, there seems to have been somewhat slower productivity growth in agriculture in Western Canada over the 1970s. Part of this slowdown occurred right at the end of the decade (1979 and 1980). Given the strong output performance of the crops sector in the prairies in 1981 and 1982, it is very likely that productivity performance has strengthened. Still, most agrologists are concerned that productivity performance in the agricultural sector is not what it could or should be.

Productivity improvement is vital if prairie producers are to retain their comparative advantage in grains and oilseeds in international markets. As well, in the early 1980s farmers have faced intensified cost-price squeeze pressures (adverse movements in the ratio of

FIGURE 8-1

AGRICULTURAL OUTPUT, INPUTS AND TOTAL FACTOR PRODUCTIVITY, WESTERN CANADA, 1961-80 (1971 = 1.000)

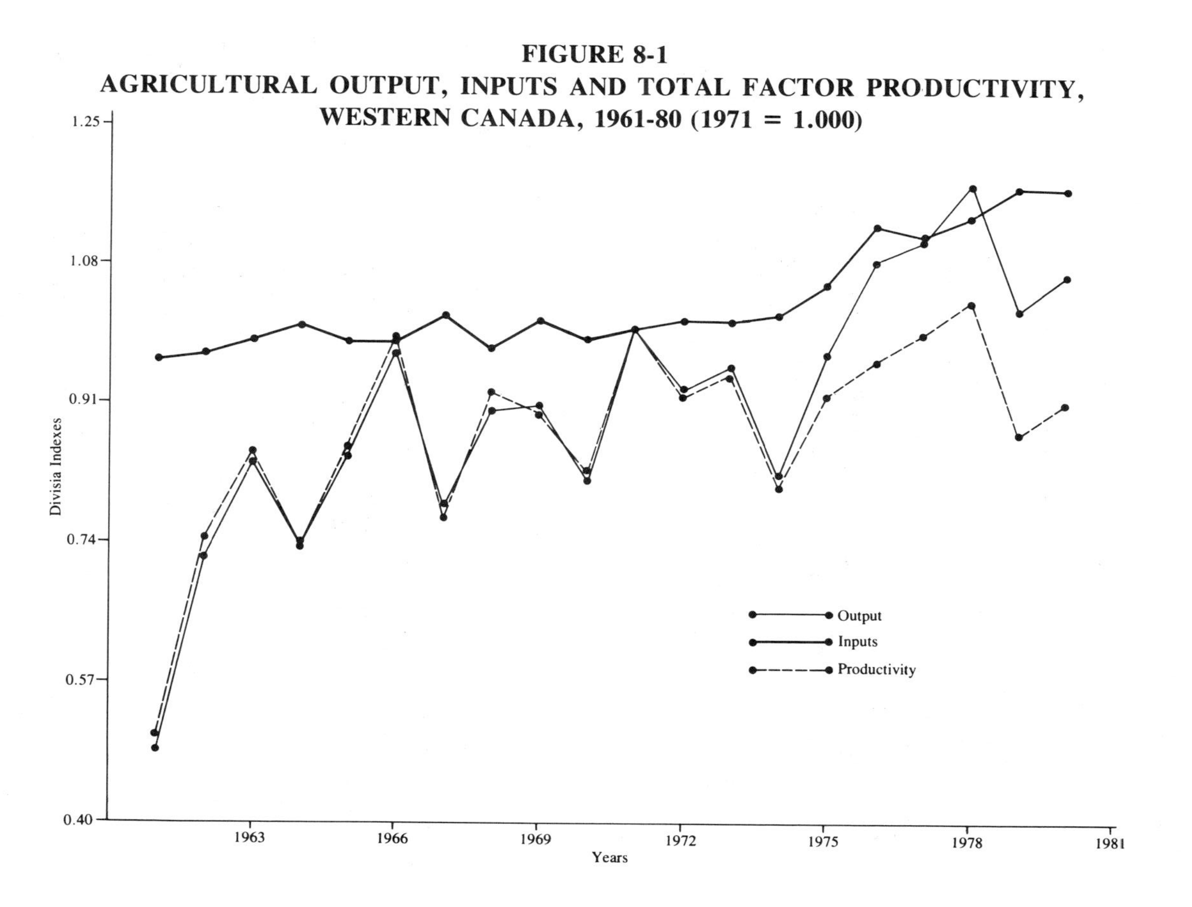

agricultural product prices to farm input prices), and productivity advance has been an historical means of alleviating such pressures.

By and large, technical innovation in the prairie grains sector since the Second World War has been more mechanical than biochemical in nature. In the future, however, biochemical aspects of innovation could be especially important. Any strategy for expanded prairie output of cereals and oilseeds must be focused on the dominant dryland sector and involve the reduction of summer fallow and the generation of enhanced dryland yields through the complementary effects of improved varieties, increased fertilizer use and improved water and snow management.[28]

The general concern with agricultural research in Canada is that it is both under-funded and under-staffed. The agricultural research institution in Canada consists of about 1,800 professionals: some 50 percent are in federal service, 35 percent in provincial institutions and universities, and 15 percent in private industry.[29] The budget of the Research Branch of Agriculture Canada is less than $200 million per year, and less than $100 million per year is spent on agricultural research at provincial and university levels. Unfortunately, at the federal level the real value of research expenditures for agriculture has eroded. The paradox is that virtually all studies on the economic returns to agricultural research — including a few Canadian studies of wheat and rapeseed — show very high returns to agricultural research. It is hard to escape the conclusion that Canada is seriously under-investing in agricultural research.

The concern with inadequate manpower for agricultural research is part of a wider problem. Most societies, including Canada, pay too little attention to the critical role of human capital formation and the quality of the labour force in agricultural development.[30] At the research professional level, there is a current need in Canada for approximately 100 new agricultural scientists per year: 80 to maintain existing research programs and 20 to accommodate moderate program expansion.[31] Better policy and planning is needed to ensure that a lack of trained manpower will not hamper agricultural research in the future. Shortages of trained professionals in such areas as plant breeding and plant pathology were being experienced in the early 1980s. At the producer level, there is a continuing need to upgrade the formal education and management skills of Canadian farmers.

Some of the problems relating to research and productivity improvement are more specific to the grains and oilseeds sector. Some fifteen years ago, Auer contended that too little emphasis was given to grains in agricultural research.[32] This criticism still seems valid, although the current imbalance in favour of horticultural crops over cereal crops is much less than that which prevailed in the mid-1960s (scc Tablc 8-1). In 1980, for example, 307 professional person-years were devoted to research on horticultural crops whereas 288 were applied to research on cereal grains and 109 to oilseeds.

Many research and productivity concerns are more crop specific. One of the historical strengths of the Canadian wheat industry has been the relatively high quality and uniformity of Canadian red spring wheat. Not only did this instill confidence in buyers but it also simplified the Canadian grain delivery system. In part, the high degree of uniformity arose because wheat breeders developed a succession of varieties, each of which in turn dominated the wheat scene (see Figure 8-2). These varieties included Red Fife at the turn of the century, Marquis from the First World War to the Great Depression, Thatcher in the decades following the Second World War, and, most recently, Neepawa, which was sown on almost 65 percent of Canada's wheat acreage in 1981.

Canada clearly wants to retain its premium market (a relatively stable market in absolute terms, but one which is a declining proportion of the total market) for high-quality, high-protein bread wheat, a market that has historically been centred in Western Europe

TABLE 8-1
DISTRIBUTION OF PROFESSIONAL PERSON YEARS (PPY) IN AGRICULTURAL RESEARCH IN CANADA BY COMMODITY/RESOURCE, 1966 AND 1980

Commodity/Resource	*PPY 1966*	*% of Total*	*PPY 1980*	*% of Total*
Forage crops	77.8	10.3	163.3	8.5
Horticultural crops	218.6	29.0	306.7	15.9
Cereal crops	112.7	15.0	288.3	14.9
Oilseed crops	14.8	2.0	109.4	5.7
Field and other crops	81.8	10.9	166.2	8.6
Animal research	222.1	29.5	454.8	23.5
Resource research	25.8	3.4	442.2	22.9
Total	753.6	100.0	1,932.1[1]	100.0

Note: [1]Includes the allocation of time for some graduate student work (to a maximum of 0.3 PPY for M.Sc. and 0.5 PPY for Ph.D.).

Source: Agriculture Canada, *Inventory of Canadian Agricultural Research* (1966 and 1980).

FIGURE 8-2
LEADING VARIETIES OF HARD RED SPRING WHEAT SOWN IN WESTERN CANADA, 1900 TO 1981

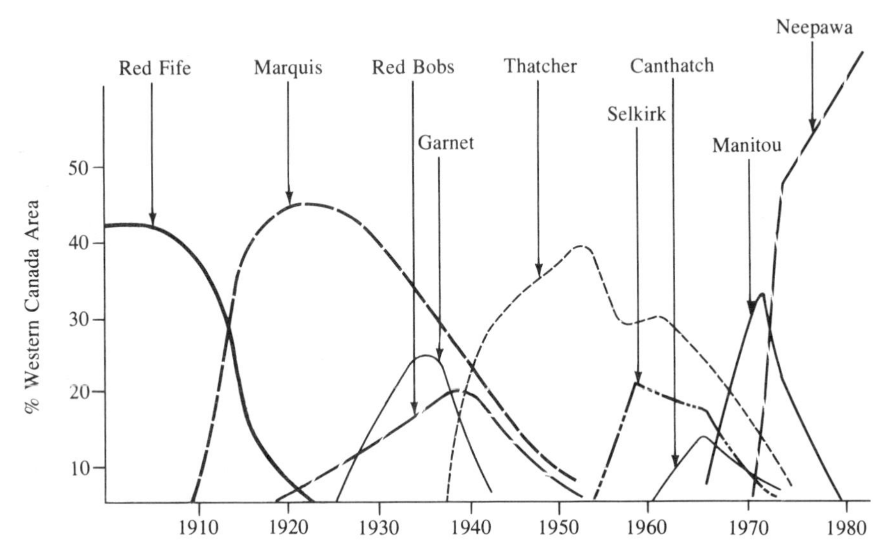

Source: W. Bushuk, "Development, Licensing, and Distribution of New Grain Varieties in Canada", in Canadian International Grains Institute, *Grains & Oilseeds*, p. 460

and Japan. It is imperative, however, that Canada expand its production and marketing base in non-traditional wheat types — soft white spring wheats, hard red winter varieties, utility wheats, and 3M (medium kernel hardness, medium protein content, and medium gluten strength) wheats — in order to capitalize on yield advantages in production and on emerging opportunities in world markets. About 20 percent of the world wheat trade consists of the hard red spring wheat class. However, the areas of growth potential in the world wheat trade appear to lie with lower quality bread wheats and wheats for other uses preferred by many developing nations. More emphasis is needed on the development of such new varieties as HY320, a 3M semi-dwarf variety of wheat, as yet unlicensed, which yields 30 percent more than Neepawa and which has acceptable quality for French bread, Arabic bread, steamed bread and noodles.[33]

The Canada Grains Council has also suggested that serious consideration should be given to lowering the minimum protein level for the licencing of hard red spring varieties.[34] A preliminary economic examination of the premium for protein indicates that it is not adequate to offset the yield advantage of lower protein varieties for many areas in Western Canada, especially the Peace River and black soil zone regions (but not southern Saskatchewan).

The requirement in Canada's grading and handling system that varieties be visually distinguishable is recognized as probably the greatest hurdle to the introduction of new and higher-yielding varieties of wheat and barley.[35] A practical and objective method of variety identification should be top priority in Canada's grain research program.[36] The challenge is to develop new higher-yielding wheat and barley varieties which are better suited for certain regions of the prairies and, at the same time, retain the advantages associated with the production and marketing of high-protein wheat and high-quality malting barley. At the very least, the economics of wheat and barley quality needs serious study. The result could well be further subregional specialization on the prairies: the growing of high protein wheat in areas like southern Saskatchewan; the growing of malting varieties in much more circumscribed areas than at present; and the growing of new higher-yielding spring wheat and feed barley varieties in areas where moisture is more assured (generally, the more northerly park-belt zone). A related problem to be considered in assessing the scope for new types and classes of wheat and barley concerns the counter-pressures to reduce the number of grades and streamline the grain handling and delivery system.

Although winter wheats yield some 20 to 25 percent more than traditional red spring varieties, research is needed to produce varieties that are sufficiently winter hardy to be grown with less risk of winter-kill in areas beyond the current small production base in southern Alberta and southwestern Saskatchewan. Soft white spring wheats, on the other hand, are best suited to irrigation areas which generate low protein levels and a product more desirable for pastry flour.[37]

All told, productivity advance in the agricultural sector — and in the grains sector specifically — does not happen accidentally but is largely the result of the political will generated for, and the societal investments made in, agricultural research. Both consumers and producers benefit from agricultural research; domestic and foreign consumers are ultimate beneficiaries, while producers can enjoy temporary income gains (short-run quasi-rents) from new technology and can maintain their foothold in domestic and international markets. The recent initiative to establish a producer check-off to generate funds for grains research and to establish a Western Grains Research Foundation is to be welcomed. But this does not obviate the need for increased public sector support for research on grains and oilseeds, given the apparent high rate of return on such investments and the public good characteristics of much research.

Land and Water Concerns

Prairie grains expansion also has policy implications with respect to the use and management of land, water and other environmental resources.[38] The most pressing land problem in Canada, and certainly in Western Canada, is not the preservation of prime agricultural land, rather it is deteriorating land quality. The role of raw land in agricultural production and development tends to be much overrated.[39] The loss of farm lands to urbanization is a rather minor factor in assessing future production levels and supply potential. What is far more crucial is the pace of productivity advance and land improvement (or decline) on the existing land base. In fact, the improved land base in the prairie provinces increased from 35.5 million hectares in 1971 to 37.7 million hectares in 1981.[40]

There is further scope for increasing the improved agricultural land base in the northern prairies. This will occur naturally, in Ricardian fashion, as either new technology or higher agricultural prices renders land at the extensive margin economically feasible for agricultural production. However, the opening of new and often marginal lands on the northern frontier should not be artificially stimulated by unwarranted subsidization or short-sighted government policy. On the other hand, it is difficult to believe that the supply of land is completely fixed or perfectly price-inelastic.

The main features of decreasing soil quality in Western Canada commonly cited include: the progressive loss of inherent fertility as some 30 to 35 percent of the original organic matter of prairie soils has been depleted; the associated loss of the natural nitrogen-supplying capacity of the soil; the deterioration in soil tilth or structure as organic matter has been lost and the consequent greater susceptibility to erosion; the increase in soil salinity, now suspected to be affecting about 10 percent of the Saskatchewan improved acreage and roughly 5 percent of the prairie acreage; and the emerging concern with soil acidification, particularly in the Peace River region of northwestern Alberta and northeastern British Columbia. Many of these soil quality problems, excluding acidification, are considered to be partly the result of the practice of summer fallowing. Summer fallowing continues to be practised on nearly 30 percent of prairie acreage. Though the share of improved land devoted to summer fallow has slowly dropped during the past decade, there seems to be scope for further reduction.

As yet, there is not a clear picture of how seriously soil quality problems are adversely affecting yields, gross returns and net farm incomes in prairie agriculture. Agricultural resource economists face a major challenge in the next decade in collaborating with soil scientists and agronomists in further assessing these problems and devising appropriate policy responses. There is reasonable agreement among prairie scientists that the deterioration in soil quality has not reached the point of irreversibility. Furthermore, there is general agreement that the productivity of prairie soils can be greatly enhanced through new strategies of land and water management in dryland agriculture, including the reduction of summer fallow, the increased use of chemical fertilizers and lime, and the more efficient use of snow and water resources. In essence, it is hoped that new methods of snow trapping and improved methods of dryland moisture conservation can generate the valuable extra inch or two of water that has been saved very inefficiently in the past by the practice of summer fallowing. Much further research needs to be done on the agronomic details of extended cropping rotations on the prairies. The Farm Lab research program in Saskatchewan is a step in the right direction.

Irrigation can make only a minor contribution to expanded grain and oilseed production, given that less than 2 percent of prairie grains and oilseeds are produced under irrigation. Furthermore, the economic merits of irrigation for traditional grain crops on the prairies are rather suspect. Grains expansion from the dryland base, as opposed to the irrigation sector, is both more socially efficient and equitable. Not only can extra grain production be achieved more cheaply through the use of society's (private and public) resources in dryland agriculture, but the benefits from expanded dryland production would be far more widely distributed among prairie producers. There are important ways, however, in which prairie grains expansion might help strengthen the economics of irrigation on the prairies. In particular, there is the possibility that the cropping mix might be intensified if a higher proportion of the irrigated acreage is devoted to soft white spring wheats, oilseeds and corn.

Grains expansion raises other environmental concerns. The possibility of more serious environmental quality problems due to increased use of agricultural chemicals and fertilizers must be monitored. Grains expansion might also involve further specialization in grains, especially wheat, and a possible movement away from a more diversified agriculture on the prairies. There are clearly questions relating to the long-run ecological sustainability of agricultural production systems in which excessive monoculture is encouraged, forage for feed or green manuring is discouraged, or a mixed crops/livestock agriculture is less likely.

Macroeconomic and Trade Policy

In the late 1970s and early 1980s, two particular facets of Canadian macroeconomic performance were adversely affecting the Canadian farm sector. First, very high interest rates hurt farmers both as producers and as debtors. Secondly, the very sluggish growth of the Canadian economy constrained growth in domestic demand for certain farm products, like red meats, which are somewhat income-responsive. Inflation, too, had impacts on Canadian agriculture.

High interest rates have raised problems in recent years for those farmers who either began or expanded operations using substantial amounts of credit. Interest costs for the farm sector in Canada nearly doubled from approximately 8 percent of total operating and depreciation expenses in the early 1970s to 14 to 15 percent in 1981 and 1982. The farm bankruptcies in 1982 and 1983, however, are indicative of severe financial problems facing a minority of farmers. The majority of farmers, already reasonably well established, have experienced some financial belt-tightening of late, but have generally enjoyed above average incomes (by historical standards) and a considerable increase in net wealth in the past decade. Nevertheless, a further movement to lower interest rates would considerably assist many producers.

Recession in the Canadian and world economies in the early 1980s has adversely affected the demand for feed grains in particular. World coarse grain exports, for example, are estimated to have declined from 104 million tonnes in 1981-82 to only 90 million tonnes in 1982-83.[41] Canada has been fortunate in having near record-level barley exports in these circumstances, although barley prices were relatively depressed. The slow (and, at times, negative) growth of real per capita income in Canada in recent years has dampened the domestic demand for beef and pork and consequently the domestic demand for feed grains. The impact of income upon demand for red meats has also been influenced by the fact that consumers have found the discretionary portion of their personal disposable incomes dropping as housing and energy-related expenditures have taken a larger share of their budgets. There are increasing signs (as suggested in Chapter 6) that there have been changes in consumer tastes and preferences away from beef. Even so, stronger growth performance in the Canadian economy would assist the red-meat sectors and indirectly stimulate feed grain use.

Since 1973 inflation has had a pervasive influence on the agricultural sector in Canada.[42] The most obvious impact, at least until 1981, was the escalation in the real prices of farmland as land buyers, primarily farmers, recognized that land was an effective hedge against inflation. Real land prices in Canada, for example, increased at 2.4 percent per year from 1962 to 1972 and then at 6.8 percent per year from 1973 to 1980. Despite the substantial capital gains of land owners over the 1970s, it is too simplistic to conclude that Canadian agriculture has benefitted from inflation. Inflation has also been associated with increased risk and uncertainty for agriculture, the intensification of cost-price squeeze pressures, and adverse impacts on the distribution of farm income and wealth. The effects of inflation on efficiency and productivity of Canadian agriculture are less clear. In the short-run, these effects may not be as deleterious as some have feared; in the long-run, the concern would be that efficiency and productivity of the sector might be more severely impaired. Inflationary pressures in the Canadian economy have abated somewhat in 1982 and 1983, but the issue of inflation will likely continue to be of some concern to Canadian agriculture.

There is also room for improvement in Canada's agricultural trade policy. Although, overall, the primary agricultural sector in Canada has a relatively low degree of protection, Canada's agricultural trade policy has slowly drifted in a more protectionist direction over the last decade. Protectionist tendencies are especially evident with the supply-managed commodities (dairy products, broilers and eggs), as well as with many fruits and vegetables and, on occasion, with meat imports. In common with nearly all advanced nations, in Canada there is an ambivalent attitude to trade which involves the advocation of less trade restrictions on commodities for which there is a comparative advantage and an increasing tendency to protect those commodities and activities for which there is not.

The grains and oilseeds sector has a strong vested interest in promoting freer agricultural trade. On this issue Canada's bargaining stance is weakened by its position on the supply-managed commodities and, indeed, by its historic commitment to the protection of the secondary manufacturing sector. Canada faces major agricultural trade barriers in the EEC and Japan, for instance. These barriers will never be easy to dismantle, but the winning of agricultural trade concessions is all the more difficult given the current federal agricultural policy that condones and further encourages the use of supply management in agriculture.

Conclusion 9

There will be renewed growth opportunities for the Western Canadian grains sector in the mid- to late 1980s, but growth prospects are more modest than was envisaged as recently as two years ago. The revised grain industry targets of 50 million tonnes of grains and oilseeds production on the prairies, with export levels of 34 million tonnes, are physically feasible and even within the realm of economic feasibility with stronger growth in the world economy. The problem is that enhanced production and export levels will not necessarily be associated with higher real grain prices nor with strongly buoyant income levels for the farm sector. World trade in grains will expand at a slower rate in the 1980s (perhaps 3 percent) than in the 1970s (7 percent). If there is any watchword for the 1980s and 1990s, it is that all actors in the grains sector — from farmers to governments — will have to be extremely flexible to take advantage of market opportunities as they arise.

The prices in world markets of grains and oilseeds have been very volatile in the last decade; instability is expected to continue. Rather than moving toward price or income support, agricultural policy should emphasize the development of an improved income stabilization policy which would enable reduction of the extent or impact of risk and uncertainty in the grains sector and which would not mask longer-run price signals. This could involve policy measures ranging from encouraging increased self-stabilization of the individual farmer (for example, through diversification of farm operations or from income tax provisions) to a strengthened (and largely farmer-financed) Western Grains Stabilization Program. The merits of moving to a total income approach to agricultural stabilization policy, as opposed to continued emphasis on individual commodity programs, warrants attention. Another question which deserves attention is the appropriate level of public funding for agriculture stabilization activities in the longer-run. Neither a national supply-management program for grains nor a world level, OPEC-style cartel for grains is likely.

If 50 million tonnes of grain were produced in 1990, the Canada Grains Council estimates that 21.3 percent would come from new lands, 48.5 percent from the reduction of the current summer fallow acreage by 2.2 million hectares, and 30.2 percent from improved yields. The relative increases in production would be similar in each of the three prairie provinces, with the potential for increase greatest in the parkland (black soil) zone. Wheat is projected to maintain its dominant share of the major crops. While it is recommended that Canada retain its high-quality, high-protein wheat market in Europe, the major market growth potential for wheat is expected to lie with wheat varieties that are higher yielding but of lower quality than the traditional hard red spring varieties. It is imperative that Canada move more swiftly to a more broadly based wheat breeding and licensing system and that non-visual means of varietal identification be developed to cope with the marketing of higher-yielding wheat and barley.

The prospects for increased use of grains within an expanded livestock sector are mixed. Recent changes in Canadian consumer preferences towards leaner cuts and lower levels of beef consumption do not favour major expansion of grain feeding by beef producers. Neither do the tendencies towards increased feed conversion efficiency and somewhat decreased emphasis on grain feeding in beef-finishing operations favour increased domestic use of feed grains. Canadian trade in beef (and the small but significant trade in beef animals) is largely with the United States. There may be prospects for modest growth in exports of beef from Western Canada to the northwestern United States. However, most international trade in beef tends to be in lower-cost and lower-priced pasture-fed meat from the southern hemisphere. The prospects for significant market expansion in poultry products are extremely limited. Even without the supply-restricting regulations on these products, substantial net exports of poultry products would be unlikely. In contrast, substantial increases in net exports of pork, primarily in fresh or frozen form, to Japan and the United States, have occurred in recent years. Export markets for pork may offer some potential for future growth in livestock product exports, although exports are unlikely to increase at as rapid a rate as in the early 1980s (when competition from Denmark, a major exporter, was limited due to disease problems).

Two areas where there have been substantial increases in grain and oilseed processing activity in recent years are oilseed crushing and the malting of barley. Canada has emerged as a major world producer and the leading world exporter of rapeseed, and rapeseed oil now comprises half the Canadian edible oil market. These achievements are largely the result of the development of improved rapeseed/canola varieties. There is potential for further expansion of economic activity in the vegetable oil milling and refining industries. However, the rapeseed crushing industry is currently suffering from excess capacity and operating losses; further public investment to encourage expansion of crushing capacity is therefore not presently warranted. The key policy issues for the rapeseed sector involve maintaining a viable crushing industry in the short-run; diversifying export markets away from heavy reliance on Japan and on raw seed exports; and continuing rapeseed research efforts to enhance yields and to produce even more acceptable oil and meal products for keenly competitive world oilseeds markets, which are currently dominated by soybeans.

The Canadian export market share of flour has tended to decline; subsidized exports from the EEC have captured an increasing share of world trade. Barley malt exports doubled over the 1970s (from a small base) and are likely to continue to increase, particularly if suitable two-row white barley varieties are introduced. Still, it must be recognized that the vast proportion of expanded grain exports — considerably more than 90 percent — would be in the form of bulk grains rather than processed grain products.

The achievement of grain export volumes of 34 million tonnes by 1990 would mean a 50 percent increase in grain handling and transportation levels over the late 1970s and a 25 percent increase over the recent record export levels. The changes in the statutory rail rates for grain will enable substantial investment to improve rail capacity. Continued efforts to improve the efficiency and productivity of the grain handling and transportation system are necessary if the benefits of potential market expansion are to accrue to the grains sector and the Canadian economy.

The potential effect of changes in the Crow rates on the relocation and encouragement of domestic grain processing activities in the prairies was given much attention in the debate that preceded the recently announced changes. However, this effect may not be as substantial as some had claimed or hoped for. There seems to have been a tendency to over-emphasize the Crow rates as a determinant of regional location decisions for some activities that are not strongly weight-losing. For stronger weight-losing activities, such as livestock feeding, interventions by some provincial governments seem likely to offset perceived improvements in the competitive position of prairie livestock producers. Moreover, the decision to pay the Crow benefit entirely to the railways, at least until 1985-86, reduces the extent of likely regional adjustments since it leaves unchanged those Crow-induced distortions among grain, livestock and related secondary processing activities, which had developed by the early 1980s. However, grain farmers will pay a share of future railway cost increases and this will limit future increases in these distortions.

Western Canadian agriculture increasingly relies on off-farm inputs, particularly farm machinery (currently comprising nearly half of farm-operating and depreciation expenses), fertilizer and agricultural chemicals. Machinery sales in the late 1970s and early 1980s have been depressed because of lower farm income, high interest rates and the fact that the current stock of farm machinery is relatively large and modern. Nevertheless, with continued grains expansion, the demand for farm machinery will gradually increase. Fertilizer use, particularly of nitrogenous fertilizers, is likely to increase steadily, as is pesticide use, particularly as farmers shift to cropping rotations involving less summer fallow. However, a large proportion of the increased pesticides used in Canadian agriculture will be imported, as will be much farm machinery, mainly from the United States.

There are limited, but important, areas where import substitution possibilities might be considered. In farm machinery, the main potential is the continuation of the efforts of smaller companies that specialize in such items as swathers and field equipment, that produce innovative products specially suited to the prairie scene, or that specialize in a very specific market segment (for example, large four-wheel drive tractors or specialized harvesting equipment). The prospects for the domestic production of agricultural chemicals are less promising, but the important candidates for further study of import-substitution are herbicides to control wild oats and 2,4-D amine weed spray. At the very least, government policy in Canada should be directed to increasing the levels of research and development expenditures by multinational chemical companies within Canada.

The contribution which the Western Canadian grains sector can make to regional and national development is contingent upon policy improvements at the federal and provincial levels in a number of areas: improved macroeconomic policy, improved trade policy, increased emphasis on agricultural research and training, enhanced investment in the skills and training of farm people, and improved policy with respect to the resource and environmental base that sustains agricultural production. The two crucial dimensions of macroeconomic policy adversely affecting the farm sector during the early 1980s have been high interest rates and the very poor growth performance of the Canadian economy.

Western Canadian agriculture has a strong interest in promoting freer agricultural trade. However, Canadian agricultural trade policy has drifted to a slightly more protectionist stance and restrictive barriers in world agricultural trade have grown in recent years. There would be significant gains to Western Canada from reductions in agricultural trade barriers.

Current expenditures on research and development for the agricultural sector are inadequate, given that improved technology is the mainspring of agricultural growth and that investment in agricultural research is regarded to have a very high payoff. Research funding for the grains sector should be expanded and grains research should be directed to the development of higher-yielding, lower-quality wheat varieties.

There is concern in agricultural circles that Canada is not training enough agricultural scientists. Improving the education and skills of farm people is also crucial in any longer-run strategy for agricultural development. Finally, policy improvements are required in land use, water use and environmental areas. The emphasis in land matters, for example, should be directed to ameliorating the deterioration of soil quality on the prairies from salinity, erosion and natural fertility losses.

In conclusion, prairie grain production and Canadian grain exports are expected to evidence moderate growth in the next decade. Production levels of 50 million tonnes and export targets of 34 million tonnes by 1990 are realistic upper-range goals and provide a useful base for current planning purposes. Grains expansion can be anticipated to strengthen gross receipts and perhaps net incomes for grains and oilseeds producers, although the dominant impact will occur through increased sales volumes, not increased price levels. Grains expansion will enable the agricultural sector to continue to run sizeable net trade surpluses and thereby make a considerable contribution to the overall current account balance. Although more than 90 percent of Canada's exports of grains and oilseeds will continue to be unprocessed, grains expansion does present modest opportunities, which should be actively pursued, to increase the amount of value-added in forward- and backward-linked industries. Taking advantage of growth opportunities in a primary resource sector such as grains and its related industries is only one part of an overall economic strategy for Canada's future.

Notes

Chapter 2: The Grains Economy

[1] Statistics Canada, *Cereals and Oilseeds Review*, Cat. 22-007 (Ottawa: Supply and Services, April 1983).

[2] Statistics Canada, *1981 Census of Canada: Agriculture*, Cat. 96-901 (Ottawa: Supply and Services, September 1982), Table 30.

[3] Canada Grains Council, *Canadian Grains Industry Statistical Handbook, 1982* (Winnipeg: Canada Grains Council), p. 39.

[4] Ibid., pp. 42-44.

[5] On average, from 1971 to 1975 oilseed products were 58 percent of the value of oilseed and product imports; this had fallen to 53 percent by 1981. See Agriculture Canada, *Canada's Trade in Agricultural Products* (Ottawa: Supply and Services, 1981 and 1982).

[6] Canadian Wheat Board, *Annual Report 1981/82* (Winnipeg: CWB, March 1983), pp. 4, 26.

[7] These data for 1981-82 are from Canada Grains Council, *Canadian Grains Industry Statistical Handbook, 1982,* pp. 170-71.

[8] The Western Producer, *Prairie Farm Policy Guide, 1981/82* (Saskatoon: The Western Producer, 1981), p. 24.

[9] For more information on the car booking system, see Grain Transportation Authority, *Grain Transportation Update* 4 (April 1983), pp. 4-5; an outline of the block shipping system is in Canadian International Grains Institute, *Grains and Oilseeds: Handling, Marketing, and Processing,* 3rd ed. (Winnipeg: CIGI, 1982), pp. 227-32.

[10] The provision for producer-loaded cars is longstanding. The number of cars for this purpose is specified by the commission. Typically, less than 1 percent of grain movement is accounted for by shipment in producer-loaded cars.

Chapter 3: International Markets

[1] Average from 1972-73 to 1981-82. Canada Grains Council, *Canadian Grains Industry Statistical Handbook, 1982* (Winnipeg: Canada Grains Council).

[2] FAO, *Trade Yearbook, 1981* (Rome: FAO, 1981).

[3] World consumption of rice exceeds that of wheat but, compared to wheat, relatively little of this food grain is traded in international markets.

[4] USDA, Economic Research Service, *Wheat: Outlook and Situation* (May 1983).

[5] USDA, Foreign Agricultural Service, *Foreign Agricultural Circular, Grains* (March 1983).

[6] As in 1963, 1965, 1972, 1975 and 1979, which were years in which the USSR's share of imports ranged from 13 to 21 percent. In the intervening years from 1960 to 1978, USSR annual imports ranged from less than 1 percent to 8.5 percent of world imports of wheat and flour. See Alan J. Webb, *World Trade in Major U.S. Crops, A Market Share Analysis,* USDA, Economics and Statistics Service (Washington: U.S. Government Printing Office, April 1981).

[7] D. Gale Johnson, *The Soviet Impact on World Grain Trade* (Washington: British-North American Committee, 1977).

[8] USDA, Economic Research Service, *Foreign Agricultural Circular, Grains* (March 1983).

[9] International Wheat Council, *Market Report* (24 February 1983).

[10] Ibid.

[11] FAO, *Trade Yearbook, 1974* and *1981*.

[12] Accounting for 21 percent and 19 percent, respectively, of world imports of wheat flour in 1981. See FAO, *Trade Yearbook, 1981*.

[13] Including some 17 percent of world exports, which were from the EEC countries under onward-processing arrangements for which destinations are not available. Canadian Wheat Board, *Grain Matters* (October 1980); and International Wheat Council, *Market Report* (24 February 1983).

[14] The market share of the United States fell in 1981-82 and in a challenge to EEC agricultural export subsidies, the United States in 1983 applied export subsidies to displace EEC export sales to Egypt, an action which led to fears of a trade war in grains.

[15] During this period, world trade in wheat and flour increased by 80 percent.

[16] FAO, *Trade Yearbook, 1981*.

[17] For an outline of intervention mechanisms for grains and oilseeds in various countries, see Cathy L. Jabara, *Trade Restrictions in International Grain and Oilseed Markets: A Comparative Country Analysis*, USDA, Economics and Statistics Service, Foreign Agricultural Economic Report No. 162 (Washington: Government Printing Office, January 1981).

[18] Canadian Wheat Board, *Grain Matters* (October 1981).

[19] Robert Bain, *Changes in the International Grain Trade in the 1980's*, USDA, Economics Research Service, Foreign Agricultural Economic Report No. 167 (Washington: Government Printing Office, July 1981), pp. 14-18. Includes a summary of opinions and studies on this issue.

[20] Walter G. Heid, Jr., *U.S. Wheat Industry,* USDA, Economics Statistics and Cooperative Service, Agricultural Economic Report No. 432 (Washington: Government Printing Office, April 1980), p. 91.

[21] A.F. McCalla, "A Duopoly Model of World Wheat Pricing," *Journal of Farm Economics* 48 (1966), pp. 711-27.

[22] C.M. Alaouze, A.S. Watson, and N.H. Sturgess, "Oligopoly Pricing in the World Wheat Market," *American Journal of Agricultural Economics* 60 (1978), pp. 173-85.

[23] C. Carter and A. Schmitz, "Import Tariffs and Price Formation in the World Wheat Market," *American Journal of Agricultural Economics* 61 (1979), pp. 517-22.

[24] C.W. Bray, P.L. Paarlberg, and F.D. Holland, *The Implications of Establishing a U.S. Wheat Board,* USDA, Economics and Statistics Service, Foreign Agricultural

Economic Report No. 163 (Washington: Government Printing Office, April 1981).

[25] A. Schmitz, A.F. McCalla, D.O. Mitchell, and C. Carter, *Grain Export Cartels* (Cambridge: Ballinger Publishing Company, 1981).

[26] USDA, Foreign Agricultural Service, *Foreign Agricultural Circular: Oilseeds and Products* (September 1982).

[27] Webb, *World Trade in Major U.S. Crops.*

[28] USDA, Foreign Agricultural Service, *Foreign Agricultural Circular: Oilseeds and Products* (February 1982).

[29] Agriculture Canada, *Market Commentary* (December 1982).

[30] Food and drug regulations in the United States restrain rapeseed oil to industrial rather than human uses, a restriction which is being appealed for the low-erucic canola varieties.

[31] USDA, *Foreign Agricultural Circular* (February 1982).

Chapter 4: Export and Production Prospects

[1] See, for example, the list of references recently compiled by Brian T. Oleson, "Long-Term Trade Prospects Outside Centrally Planned Countries," Agriculture Canada, *Market Commentary: Proceedings of the Canadian Agricultural Outlook Conference* (December 1982), pp. 82-89.

[2] Anthony Rojko et al., *Alternative Futures for World Food in 1985*, Volume 1, *World GOL Model Analytical Report*, USDA, Economics Statistics and Cooperative Service, Foreign Agricultural Economics Report (Washington: Government Printing Office, 1978).

[3] International Food Policy Research Institute, *Food Needs of Developing Countries: Projections of Production and Consumption to 1990* (Washington: IFPRI, 1977).

[4] As reported in *The Western Producer,* 18 August 1983.

[5] See the most recent estimates of the Population Reference Bureau, *1983 World Population Data Sheet* (Washington: Population Reference Bureau, April 1983). These PRB estimates contain the results of the 1982 census in China in which the Chinese population was estimated at slightly more than 1 billion people and growing at a (slightly higher than anticipated) rate of 1.5 percent per year. As a consequence, the PRB estimate of the rate of world population increase was raised to 1.8 percent from the 1.7 percent figure cited in 1982 and 1981.

[6] For a more detailed and critical evaluation of the world food situation, see D. Gale Johnson, "The World Food Situation: Developments During the 1970s and Prospects for the 1980s," in Emery N. Castle and Kenzo Hemmi, eds., *U.S.-Japanese Agricultural Trade Relations* (Baltimore: The Johns Hopkins University Press for Resources for the Future, 1982), pp. 15-57.

[7] In assessing the long-term future production potential of Canadian or American agriculture, we share the guardedly optimistic outlook outlined in a recent overview article; see Sandra S. Batie and Robert G. Healy, "The Future of American Agriculture," *Scientific American* 248 (February 1983), pp. 45-53.

[8] U.S., Council on Environmental Quality and the Department of State, *The Global 2000 Report to the President,* 3 vols. (Washington: Government Printing Office, 1980).

[9] Agriculture Canada, *Challenge for Growth: An Agri-Food Strategy for Canada,* Discussion Paper (Ottawa: Minister of Agriculture, July 1981).

[10] G.O. Barney, P.H. Freeman, and C.A. Ulinski, *Global 2000: Implications for Canada* (Toronto: Pergamon Press, 1981).

[11] Canada Grains Council, *Prospects for the Prairie Grain Industry, 1990* (Winnipeg: Canada Grains Council, November 1982).
[12] As reported in *The Western Producer,* 18 August 1983.
[13] Organization for Economic Co-operation and Development, *Prospects for Soviet Agricultural Production and Trade* (Paris: OECD, 1983).
[14] International Wheat Council, *Market Report,* Special Section on China: Grain Import Prospects (27 October 1983).
[15] Canada Grains Council, *Prospects for the Prairie Grain Industry 1990.*

Chapter 5: Economic Impact

[1] See, for example, Vernon C. Fowke, *The National Policy and the Wheat Economy* (Toronto: University of Toronto Press, 1957); and D. Owram, *The Economic Development of Western Canada: An Historical Overview*, Economic Council of Canada, Discussion Paper No. 219 (Ottawa: ECC, November 1982).
[2] Fowke, *The National Economy and the Wheat Economy*, pp. 72-73.
[3] Quoted in *Canadian Annual Review of Public Affairs* (1905), pp. 149-50; reprinted in Fowke, *The National Policy and the Wheat Economy,* p. 66; and in Owram, *The Economic Development of Western Canada,* p. 12.
[4] John Richards and Larry Pratt, *Prairie Capitalism: Power and Influence in the New West* (Toronto: McClelland and Stewart, 1979), p. 314.
[5] Vernon C. Fowke, *Canadian Agricultural Policy* (Toronto: University of Toronto Press, 1946), pp. 280-81.
[6] Owram, "The Economic Development of Western Canada," pp. 20-22 (as derived from the 1931 Census of Canada).
[7] Calculated from data in Statistics Canada, *Provincial Gross Domestic Product by Industry, 1980,* Cat. 61-202 (Ottawa: Supply and Services, April 1983). The limited service sectors covered are education, hospitals, and accommodation and food services.
[8] Canada, Royal Commission on Canada's Economic Prospects, *Progress and Prospects of Canadian Agriculture* (Ottawa: Queen's Printer, 1957), p. 13.
[9] Terrence S. Veeman and Michele M. Veeman, "The Changing Organization, Structure, and Control of Canadian Agriculture," *American Journal of Agricultural Economics,* Proceedings Issue (December 1978), pp. 759-68.
[10] S.E. Drugge and T.S. Veeman, "Industrial Diversification in Alberta," *Canadian Public Policy* 6 Supplement (1980), pp. 221-28.
[11] Canada Grains Council, *Prospects for the Prairie Grain Industry 1990* (Winnipeg: Canada Grains Council, November 1982).
[12] Ibid., pp. 190-201.
[13] *Globe and Mail,* 2 August 1983.
[14] Statistics Canada, *National Income and Expenditure Accounts,* various years (Ottawa: Supply and Services).
[15] Agriculture Canada, Regional Development Branch, *Selected Agricultural Statistics: Canada and the Provinces, 1983 (Ottawa: Supply and Services, June 1983), p. 62.*
[16] Statistics Canada, *The Input-Output Structure of the Canadian Economy 1971-79,* Cat. 15-201E (Ottawa: Supply and Services, May 1983) (1979 Impact Matrix; Aggregation M).
[17] Canada Grains Council, *The Role and Importance of Grains and Oilseeds to the Canadian Economy* (Winnipeg: Canada Grains Council, January 1983).

[18] Richard Caves and Richard Holton, *The Canadian Economy: Prospect and Retrospect* (Cambridge: Harvard University Press, 1959), p. 215; and quoted in Owram, "The Economic Development of Western Canada," p. 40.
[19] Owram, p. 40.
[20] Drugge and Veeman, "Industrial Diversification in Alberta," pp. 221-23.

Chapter 6: Processing and Use of Canadian Grains

[1] Statistics Canada, *Supply and Disposition of Major Grains, Canada* (Ottawa: Statistics Canada, May 1983); and Canada Grains Council, *Canadian Grains Industry Statistical Handbook, 1982* (Winnipeg: Canada Grains Council).
[2] For documentation and subsequent references to consumption data from 1960 to 1980, see Agriculture Canada, Marketing and Economics Branch, *Handbook of Food Expenditures Prices and Consumption,* Publication No. 8115 (Ottawa: Supply and Services, 1981); later per capita consumption data were taken from Agriculture Canada, *Market Commentary* (June 1983), p. 25; and Statistics Canada, *Apparent Per Capita Food Consumption in Canada* (Ottawa: Supply and Services).
[3] These utilization estimates are from John R. Groenewegen, "The Canadian Coarse Grains Industry," Agriculture Canada, Working Paper, June 1982.
[4] Statistics Canada, *Cereals and Oilseeds Review* (Ottawa: Supply and Services, April 1983).
[5] Groenewegen, "The Canadian Coarse Grains Industry."
[6] See, for example, Zuhair A. Hassan and S.R. Johnson, *Consumer Demand for Major Foods in Canada* (Ottawa: Agriculture Canada, Economics Branch 1976).
[7] Agriculture Canada, Marketing and Economics Branch, *Canada's Trade in Agricultural Products: 1980, 1981, and 1982,* Publication No. 83/4 (Ottawa: Supply and Services, September 1983).
[8] Canadian Wheat Board, *Annual Report 1981/82* (Winnipeg: CWB, 1982), p. 16.
[9] In addition, the two grades of wheat for which the previous CWB losses in export revenues had been the largest — No. 3CWRS and No. 1 Canada Utility — were removed from the corn competitive pricing system.
[10] The Canadian Grains Commission, among its other activities, previously specified and monitored the maximum levels of elevator space for non-board feed grains. These restrictions were discontinued in 1982.
[11] Livestock Feed Board of Canada, *Annual Report, 1981/82* (Ottawa: Livestock Feed Board, 1982), p. 20.
[12] For a full listing of qualifying products, see J.C. Gilson, *Western Grain Transportation: Report on Consultations and Recommendations* (Ottawa: Supply and Services, 1982), p. V-6.
[13] See Canola Crushers of Western Canada, *A State of the Industry Report on Canola Crushing in Western Canada* (Winnipeg: Canola Crushers of Western Canada, June 1982), pp. 27-31.
[14] For example, see the criticism of the most extensive of the published studies of the Crow rates, written by D.R. Harvey and cited below, in the review by Gillian Wogin, in *Canadian Journal of Economics* 15 (August 1982), pp. 555-57.
[15] D.R. Harvey, *Christmas Turkey or Prairie Vulture: An Analysis of the Crow's Nest Pass Rates* (Montreal: The Institute of Research on Public Policy, 1980).
[16] Statistics Canada, *Cereals and Oilseeds Review* (Ottawa: Supply and Services, April 1983).

[17] Canadian Wheat Board, *Grain Matters* (November/December 1981).
[18] Statistics Canada, *Flour and Breakfast Cereal Products Industry, 1981*, Cat. 32-228 (Ottawa: Supply and Services, 1983).
[19] Canada Grains Council, *Canadian Grains Industry Statistical Handbook, 1982.*
[20] FAO, *Trade Yearbook, 1981* (Rome: FAO, 1981).
[21] Canadian International Grains Institute, *Grains and Oilseeds,* 3rd ed. (Winnipeg: CIGI, 1982).
[22] Statistics Canada, *Supply and Disposition of Major Grains.*
[23] Statistics Canada, *Oils and Fats,* Cat. 32-006 (Ottawa: Supply and Services, December 1982).
[24] For a listing, see Statistics Canada, *Oils and Fats.*
[25] For example, one-fifth of the capital costs of Canada Packers' recently completed rapeseed crushing plant in Hamilton was financed by an Ontario government grant. See *Globe and Mail*, 20 June 1983.
[26] Canada Grains Council, *Prospects for the Prairie Grain Industry, 1990.*
[27] Grain Transportation Authority, *Grain Transportation Update* 4 (April 1983).
[28] Grain Transportation Authority, *Grain Transportation Update* 3 (August 1982), pp. 3-4.

Chapter 7: Input Supply Industries

[1] Canada, Royal Commission on Farm Machinery, *Report of the Royal Commission* (Ottawa: Information Canada, 1971), Chapter 7.
[2] Clarence Barber, "The Farm Machinery Industry: Reconciling the Interests of the Farmer, the Industry, and the General Public," *American Journal of Agricultural Economics 55* (December 1973), pp. 820-28.
[3] Archie N. Book, "Farm Machinery Retailing in Canada," *Canadian Farm Economics* (June 1979), pp. 25-37.
[4] *Report of the Royal Commission on Farm Machinery*, Chapter 13.
[5] *The Western Producer,* 14 April 1983, p. A53.
[6] *The Western Producer,* 21 April 1983, p. A9.
[7] *Globe and Mail,* 14 June 1983, p. B15.
[8] Statistics Canada, Manufacturing and Primary Industries Division, "Concentration and Foreign Control in Provincial Markets, Manufacturing Industries, 1979" (Ottawa: Supply and Services, 1979).
[9] Statistics Canada, *Agricultural Implement Industry* (Ottawa: Supply and Services, annual).
[10] A.N. Book and P. Kampouris, *Statistics Relating to Farm Machinery in Canada 1950 to 1976,* Agriculture Canada, Economics Branch (Ottawa: Supply and Services, July 1977); and *Statistics Canada Daily*, 26 June 1983, p. 5.
[11] Statistics Canada, Cat. 42-202; and Book and Kampouris, *Statistics Relating to Farm Machinery.*
[12] *Globe and Mail,* 29 November 1982, p. B1.
[13] W.G. Phillips, *The Agricultural Implement Industry in Canada: A Study of Competition,* Canadian Studies in Economics No. 7 (Toronto: University of Toronto Press, 1956), p. 79.
[14] Statistics Canada, *Exports, Merchandise Trade* (Ottawa: Supply and Services); and Statistics Canada, *Imports, Merchandise Trade* (Ottawa: Supply and Services).
[15] Ian F. Furniss, "Fertilizers," Agriculture Canada, *Market Commentary: Farm Inputs and Finance* (December 1982), pp. 69-75.

[16] Ibid.
[17] Calculated from Z. Piracha, "Fertilizers," Agriculture Canada, *Market Commentary, Farm Inputs and Finance* (December 1981), p. 16.
[18] R.F. Leibenluft, *Competition in Farm Inputs: An Examination of Four Industries,* Policy Planning Issues Paper, U.S. Federal Trade Commission, Office of Policy Planning, Report No. FTC/OPP-81-05 (Washington: U.S. Department of Commerce, National Technical Information Service, February 1981), p. 34.
[19] Leibenluft, *Competition in Farm Inputs*, pp. 36-37.
[20] Agriculture Canada, Regional Development Branch, *Selected Agricultural Statistics, Canada and the Provinces* (Ottawa: Supply and Services, 1983), pp. 27-32.
[21] Ibid., p. 31.
[22] Agriculture Canada, *Orientation of Canadian Agriculture*, Vol. 1, Part A (Ottawa: Supply and Services, 1977), p. 129.
[23] D.A. Rennie, J.D. Beaton, and R.A. Hedlin, *The Role of Fertilizer Nutrients in Western Canadian Development* (Calgary: Canada West Foundation, 1980).
[24] Furniss, "Fertilizers," p. 80.
[25] Rennie, *"The Role of Fertilizer Nutrients."*
[26] Furniss, "Fertilizers," p. 80.
[27] W.H. Horner et al., *Western Canadian Agriculture to 1990* (Calgary: Canada West Foundation, 1980), pp. 188-89.
[28] E. Suen and Z. Piracha, "Agricultural Chemicals," Agriculture Canada, *Market Commentary, Farm Inputs and Finance* (December 1982), p. 12.
[29] V.A. Heighton, "Agricultural Chemicals and Other Supplies," Agriculture Canada, *Market Commentary, Farm Inputs and Finance* (December 1981), p. 22.
[30] Horner, *Western Canadian Agriculture to 1990,* p. 199.
[31] R. Hay, *Pesticides and Herbicides: Present and Anticipated Use by 1990* (Calgary: Canada West Foundation, 1980).
[32] Horner, *Western Canadian Agriculture to 1990,* p. 200.
[33] Agriculture Canada, *Orientation of Canadian Agriculture*, pp. 133-34.
[34] T.R. Eichers, *The Farm Pesticide Industry,* USDA, Economics Statistics and Cooperative Service, Agricultural Economic Report No. 461 (Washington: Government Printing Office, September 1980), pp. 7-12.
[35] Leibenluft, *Competition in Farm Inputs*, p. 55.
[36] Hay, *Pesticides and Herbicides*, p. 14.
[37] Agriculture Canada, *Orientation of Canadian Agriculture*, p. 135.
[38] M.H. Hawkins, R.R. Norby, and J.H. Copeland, "A 'Made-in-Canada' Agricultural Chemical Industry," Paper presented to the 1980 Annual Conference of the Canadian Agricultural Chemical Association, September 1980, p. 20.
[39] R. Krystynak, "An Economic Assessment of 2,4-D in Canada: The Case of Grain," *Canadian Farm Economics* 18, p. 24.
[40] Agriculture Canada, *Orientation of Canadian Agriculture*, p. 135.
[41] Canada, Industry, Trade and Commerce, Chemicals Branch, "An Analysis of the Pesticides Industry in Canada," Unpublished discussion paper, 1982.
[42] Agriculture Canada, *Orientation of Canadian Agriculture*, p. 133.
[43] Twenty-one of the thirty-six member companies of the Canadian Agricultural Chemical Association are subsidiaries of the U.S. active ingredient producers listed by Eicher.
[44] Agriculture Canada, *Orientation of Canadian Agriculture*, p. 134.
[45] Agriculture Canada, *Selected Agricultural Statistics*, p. 32.

[46] Canada, Industry, Trade and Commerce, Chemicals Branch, "An Analysis of the Pesticides Industry in Canada," pp. 5, 35.

Chapter 8: Policy Issues and Constraints

[1] Carl Snavely, *Summary of Report on 1980 Costs and Revenues Incurred by the Railways in the Transportation of Grain Under the Crowsnest Pass Rates* (Ottawa: Minister of Transport, 1982).

[2] Involving subsidies to compensate for uneconomic branch lines and the capitalized portion of expenditures to rehabilitate branch lines, purchase grain hopper cars and repair boxcars.

[3] J.C. Gilson, *Western Grain Transportation: Report on Consultations and Recommendations* (Ottawa: Supply and Services, 1982), p. II-7.

[4] The Canadian Wheat Board, *Annual Report, 1977/78* (Winnipeg: CWB, 1978).

[5] D.W. Gillen and T.H. Oum, "Railways in Western Canada: Bottlenecks, Capacity Expansion and Financing," Paper submitted to the Economic Council of Canada, n.d.

[6] The Crow benefit is specified as a smaller sum for the years preceding 1986-87.

[7] The base period quantity is specified as 31.1 million tonnes.

[8] This is more fully discussed by K.H. Norrie, "Not Much to Crow About: A Primer on the Statutory Grain Freight Issue," *Canadian Public Policy* (forthcoming).

[9] H. Bruce Huff, "Mid-term Outlook for Canadian Agriculture," *Market Commentary, Proceedings of the Canadian Agricultural Outlook Conference* (December 1982) p. 44.

[10] Unpublished research by one of the authors.

[11] Based on comparison of insurance participants — Horner, *Western Canadian Agriculture to 1990* (Calgary: Canada West Foundation, 1980) — with CWB permit-holder numbers — Canadian Wheat Board, *Annual Report, 1981/82* (Winnipeg: CWB, 1982).

[12] Canada, Federal-Provincial Relations Office, *Federal-Provincial Programs and Activities* (Ottawa: Supply and Services, March 1982), p. 4.

[13] For a further description, see Charles F. Wilson, *Grain Marketing in Canada* (Winnipeg: Canadian International Grains Institute, 1979), pp. 273-83.

[14] Agriculture Canada, *Western Grain Stabilization, Annual Report, 1980* (Ottawa: Minister Responsible for the Western Grains Stabilization Act and Minister of Agriculture, 1982).

[15] Cited in Colin A. Carter, *The System of Marketing Grain in Canada,* Extension Bulletin No. 82-2 (Winnipeg: University of Manitoba, Department of Agricultural Economics and Farm Management, July 1982), p. 29.

[16] M.E. Glenn, C.A. Carter, and O.P. Tangri, "Government Support in the Grain Sector: A Canadian-U.S. Comparison," Working Paper, University of Manitoba, Department of Agricultural Economics, January 1983.

[17] I. Roberts and G. Tie, "The Emergence of the EEC as a Net Exporter of Grain," *Quarterly Review of the Rural Economy 4* (Australia: Bureau of Agricultural Economics, November 1982), pp. 295-304.

[18] Canadian Wheat Board, "Brief to the House of Commons Standing Committee on Transport" (Winnipeg: Canadian Wheat Board, 1983), pp. 14-20.

[19] See, for example, Dan Morgan, *Merchants of Grain* (New York: The Viking Press, 1979).

[20] David L. Watt, Donald O. Mitchell, and John Ross, "A Comparison of the Current United States and Canadian Wheat Marketing Systems," Agricultural Economics Staff Paper No. 79-22 (Ann Arbor: Michigan State University, 1979).

[21] Alex F. McCalla and Andrew Schmitz, "Grain Marketing Systems: The Case of the United States versus Canada," *American Journal of Agricultural Economics* 61 (May 1979), pp. 199-212.

[22] Foodwest Resource Consultants, *U.S. Grain Handling and Transportation With Selected Comparisons to the Canadian System* (Edmonton: Alberta Transportation, 1979).

[23] Canadian Wheat Board, *Grain Matters* (November 1979), pp. 4-7.

[24] Keith Peltier and Donald E. Anderson, *The Canadian Grain Marketing System,* Agricultural Economics Report No. 130 (Fargo: North Dakota State University, Department of Agricultural Economics, December 1978).

[25] Canadian Wheat Board, *Grain Matters* (November 1979), p. 5.

[26] D.R. Harvey, *Government Intervention and Regulation in the Canadian Grains Industry,* Technical Report E/16 (Ottawa: Economic Council of Canada and The Institute for Research on Public Policy, 1981), pp. 79-87.

[27] These estimates involve an update and refinement of the initial research work presented in Tariq S. Islam, "Changing Input Use and Productivity in Canadian Agriculture," Unpublished Ph.D. dissertation, University of Alberta, 1982.

[28] See T.S. Veeman, "The Role of Water Management in the Expansion of Prairie Grain Production: Economics and Policies," in Canadian Wheat Board, Advisory Committee, *Prairie Production Symposium,* Saskatoon, October 1980 (Winnipeg: CWB, 1980).

[29] A.A. Guitard, "The Canadian Agricultural Research Institution," Address to the Conference on Economics of Agricultural Research in Canada, Lethbridge, Alberta, 8 September 1983.

[30] See, for example, T.W. Schultz, "The Economics of Being Poor," *Journal of Political Economy* 88 (August 1980), pp. 639-51.

[31] A.A. Guitard, "The Ph.D. Shortage," *Agrologist* (Winter 1982), pp. 14-15.

[32] L. Auer, *Canadian Agricultural Productivity,* Economic Council of Canada Staff Study No. 24 (Ottawa: Economic Council of Canada, 1969).

[33] R. DePauw, "Future Wheat," *Irrigation Outlook* (Saskatchewan Agriculture, April 1983), pp. 8-9.

[34] Canada Grains Council, *Grain Grading for Efficiency and Profit* (Winnipeg: Canada Grains Council, September 1982).

[35] See the excellent overview by W. Bushuk, "Plant Science-Summation," Presentation to the Prairie Production Symposium sponsored by Canadian Wheat Board, Saskatoon, October 1980.

[36] Ibid., p. 16.

[37] See Canada Grains Council, *The Soft White Spring Wheat Industry in Canada* (Winnipeg: Canada Grains Council, July 1982).

[38] The discussion in this section rests heavily on T.S. Veeman, "Maintaining the Quantity and Quality of the Agricultural Land Base," in *Proceedings of the 1981 Annual Meeting of the Canadian Agricultural Economics Society*, Brock University, 9-13 August 1981 (October 1982), pp. 70-101; and T.S. Veeman, "The Role of Water Management in the Expansion of Prairie Grain Production: Economics and Policies."

[39] Schultz, "The Economics of Being Poor."

[40] Statistics Canada, *Census of Canada 1981* and *1971: Agriculture* (Ottawa: Supply and Services).

[41] Agriculture Canada, *Market Commentary* (June 1983), p. 13.

[42] M.M. Veeman and T.S. Veeman, ''Inflation and Canadian Agriculture: Effects on Efficiency, Productivity, and Equity,'' in International Association of Agricultural Economists, *Growth and Equity in Agricultural Development,* Contributed papers read at the 18th International Conference of Agricultural Economists, Jakarta, August 1982, IAAE Occasional Paper No. 3 (forthcoming).

The Canadian Institute for Economic Policy Series

The Monetarist Counter-Revolution: A Critique of Canadian Monetary Policy 1975-1979
Arthur W. Donner and Douglas D. Peters

Canada's Crippled Dollar: An Analysis of International Trade and Our Troubled Balance of Payments
H. Lukin Robinson

Unemployment and Inflation: The Canadian Experience
Clarence L. Barber and John C.P. McCallum

How Ottawa Decides: Planning and Industrial Policy-Making 1968-1980
Richard D. French

Energy and Industry: The Potential of Energy Development Projects for Canadian Industry in the Eighties
Barry Beale

The Energy Squeeze: Canadian Policies for Survival
Bruce F. Willson

The Post-Keynesian Debate: A Review of Three Recent Canadian Contributions
Myron J. Gordon

Water: The Emerging Crisis in Canada
Harold D. Foster and W.R. Derrick Sewell

The Working Poor: Wage Earners and the Failure of Income Security Policies
David P. Ross

Beyond the Monetarists: Post-Keynesian Alternatives to Rampant Inflation, Low Growth and High Unemployment
Edited by David Crane

The Splintered Market: Barriers to Interprovincial Trade in Canadian Agriculture
R.E. Haack, D.R. Hughes and R.G. Shapiro

The Drug Industry: A Case Study of the Effects of Foreign Control on the Canadian Economy
Myron J. Gordon and David J. Fowler

The New Protectionism: Non-Tariff Barriers and Their Effects on Canada
Fred Lazar

Industrial Development and the Atlantic Fishery: Opportunities for Manufacturing and Skilled Workers in the 1980s
Donald J. Patton

Canada's Population Outlook: Demographic Futures and Economic Challenges
David K. Foot

Financing the Future: Canada's Capital Markets in the Eighties
Arthur W. Donner

Controlling Inflation: Learning from Experience in Canada, Europe, and Japan
Clarence J. Barber and John C.P. McCallum

Canada and the Reagan Challenge: Crisis in the Canadian-American Relationship
Stephen Clarkson

The Future of Canada's Auto Industry: The Big Three and the Japanese Challenge
Ross Perry

Canadian Manufacturing: A Study of Productivity and Technological Change
Volume I: Sector Performance and Industrial Strategy
Volume II: Industry Studies 1946-1977
Uri Zohar

Canada's Cultural Industries: Broadcasting, Publishing, Records and Film
Paul Audley

Canada's Video Revolution: Home Video, Pay-TV and Beyond
Peter Lyman

Marketing Canada's Energy: A Strategy for Security in Oil and Gas
I.A. McDougall

Offshore Oil: Opportunities for Industrial Development and Job Creation
Roger Voyer

Planning and the Economy: Building Federal-Provincial Consensus
H.G. Thorburn

Economic Recovery for Canada: A Policy Framework
John Cornwall and Wendy Maclean

Business Cycles in Canada: The Postwar Experience and Policy Directions
Maurice Lamontagne

World Economy in Crisis: Unemployment, Inflation and International Debt
Lorie Tarshis

The above titles are available from:
James Lorimer & Company, Publishers
Egerton Ryerson Memorial Building
35 Britain Street
Toronto, Ontario M5A 1R7